Ric Natt

Talking W

Acknowledgements

Talking Wildlife really began back in 1993 when I was invited by ABC radio producer Louise McCosker to come into the studio and take calls from the listeners. Louise produced the Jan Taylor program on 612 4QR Brisbane which ran weekdays from 10 a.m. till midday. I will be forever grateful to Louise and Jan and all the producers and presenters over the past 12 years. They are Andrew Lofthouse, Peter Gooch, Haydn Sargent, Anna Reynolds, Carolyn Tucker, Tracey Strong, Cathy Border, Kelly Higgins-Divine, Ann Debert, Pam O'Brien, Zara Horner, Dennis Martin, Chris Welch and all the other ABC producers and broadcasters who have stood in for temporarily absent colleagues and who have always been so kind and patient with me.

I thank Dr Greg Gordon of the Queensland Parks and Wildlife Service who first offered me the position of animal attendant in that organisation and who has now had more than 20 years to review the wisdom of that decision. I also thank all my other former QPWS colleagues who shared their knowledge and learning and helped me to understand the principles of scientific publication.

I am especially indebted to Mr Paul Grimshaw of the QPWS who taught me just about everything I know about birds and their calls and what little I still know of plants. I thank Paul also for his friendship and the many hours we've spent together in the bush.

Griffith University's Dr Darryl Jones is owed a special debt as well for the encouragement he has given me for this book and for his dedication to the science so desperately needed in the uncertain world of wildlife management.

I thank my wife Chris Boston who read all the early drafts and told me which bits I'd left out. And finally I am grateful to Peter Slater who suggested Wildlife Talkback as a book, and to Steve Parish Publishing, particularly Ann Wright, Kate Lovett, Cristina Pecetta, Emma Ralston and Britt Winter, for without them this would just be a vague idea.

First published in Australia by
Steve Parish Publishing Pty Ltd
PO Box 1058, Archerfield, Queensland 4108 Australia

www.steveparish.com.au

Produced at the Steve Parish Publishing Studios, Australia

ISBN 174021540 0

Photography: Ric Nattrass (uncredited photographs), Alan and Stacey Franks, Paul Grimshaw, Ian Morris, Emma Ralston and Steve Parish, as credited.

Design by Cristina Pecetta, SPP
Printed by Printplus Ltd, Hong Kong
Film by Colour Chiefs Digital Imaging, Brisbane

Foreword

The only real enemy of talkback radio is time. For it to work well, a talkback program needs to be crisp and very much to the point. With a topic as broad and important as "wildlife", there are inevitably calls that would benefit from much greater detail and far longer explanations. However, there is a delicate balance between retaining a large and interested audience and giving enough time and detail to each call.

For the past twelve years Ric Nattrass has demonstrated an extraordinary natural ability to balance the two, but he is the first to admit that there are many occasions when detail is sacrificed to the tyranny of time.

Fortunately, what you can't do on talkback you can do in a book, and Talking Wildlife is it. Here, Ric addresses some of the most popular questions and topics, discusses them at length, without the nagging, ticking clock and yet in the same witty and sometimes hilarious style that has made him a great favourite on radio, at the Greenhouse at the Woodford Folk Festival and as an after dinner speaker.

Whether you've ever experienced Dux of the Pool or Turkish Delight personally, you could develop a more sympathetic view of the birds' plight, and there is likely to be some hair-tearing and teeth-gnashing in more traditional circles when To Feed or Not to Feed is only partly digested.

I know from personal experience there is a lot more inside the head of this extraordinary former Wildlife Ranger, so don't be surprised if in due course you see Talking Wildlife two, three or even four.

Sandy McCutcheon
Brisbane 2004

Contents

Note: "Received" common names, such as those recorded in the definitive zoological texts, are capitalised. Other common names are given lower case.

Introduction

Twelve years ago (is it really that long?) I was invited to the ABC to sit in for the vet who had taken a few weeks holiday. The two subjects are only distantly related, but the theme was animals and the producer thought I had enough experience in my chosen field to be unleashed on the listeners. For the next couple of years, the presenters and producers of the 10 a.m. to midday program alternated vet talkback with wildlife talkback each week until the popularity of the wildlife calls took over.

Some people have tried to blame me entirely for the success of Wildlife Talkback but this would be grossly unfair. Firstly, people are primarily listening to the regular presenter and guests who drop in for 30 minutes to take calls on their special subject have it easy.

In my case there have been extraordinarily popular presenters produced by some very astute people and, in any event, the popularity of wildlife had, by 1993, been a burgeoning public interest for at least a decade. It is the wildlife that is responsible for the segment's success, that and the many ways it is perceived by the callers.

I always cringe at the label "expert". Not out of false modesty either. Nothing could be further from the truth. It is hugely flattering, but it is simply not true. An expert knows an awful lot of detail about his/her special subject and "wildlife" is just too big a subject to be an expert in. I'm a generalist who has had the great fortune to be educated, not just by Nature itself, but by the thousands of callers who have brought to my attention a lot of things I would never have had time to see for myself.

Wildlife Talkback is a case of flying by the seat of your pants. I'm not literate enough to have a laptop computer with a whole range of data on it that I can look up if I'm stumped. I learned very early in the piece that I either knew or I didn't and the only thing to do when I didn't was to just say so and then go and find out. That becomes "homework". Since the very beginning, I just waltz in with my diary, with the numbers 1 to 12 written down the side of the page marked Friday the whatever, under my wing, pull the headphones on and hope for the best. At the end of the session, which is always the fastest 25 minutes in any week, the result can look something like this...

1.	*John*	*Townsville*	*Bird ID (Crested Hawk)*
2.	*Maryanne*	*Caboolture*	*Noise at night (brushtail possum)*
3.	*Bobbie*	*Taringa*	*Crows and Ringtails*
4.	*Catherine*	*Albany Creek*	*Dingo*
5.	*John*	*Gympie*	*How to keep ducks out of pool*
6.	*Jean*	*Bray Park*	*Sick looking cockatoo*
7.	*Nancy*	*Cairns*	*Bird book (which one)*
8.	*David*	*Peregian Beach*	*Bird ID (Figbird)*
9.	*Harold*	*Rocky*	*Frog ID (Graceful Tree-frog)*
10.	*Bev*	*Redlands*	*Hairy caterpillars (Ochrogaster lunifer)*
11.	*Jean*	*Ashgrove*	*Spider (Golden Orb-weaver)*
12.	*Barry*	*Logan*	*Carpet Python in dog's water bowl*

I gathered up some of my old diaries and loaded these data into a bit of a computer program to get some idea of what the pattern of calls looked like. Perhaps not surprisingly birds dominate at 48% of all calls. Reptiles are runners up but a fair way back with 18%, mammals 14%, and all other categories like frogs and insects are less than 1% each.

Abusive or "negative" calls are fairly rare and I've often wondered whether this is a reflection of people's general view of the natural world or whether it says more about the characteristics of ABC listeners. I like to think it might be equal proportions of both. Most weeks there is a small pile of letters from callers who have chosen to write rather than call. The most common reasons for the letters include "couldn't get through", "I'm at work when your show is on", "don't have the courage" (and I know how they feel – I'm the same!) and the fact that some listeners don't feel that the limited time on the air is sufficient to allow a full description of their question. Many of these letters have also included photos of bird antics, odd-bods like partial albinism and hybrids like Galah/Long-billed Corella crosses. It has become a wonderful collection and I feel privileged to be its custodian. Thank you to all those listeners who have taken the time and trouble to contribute to this collection.

On air, there is just no time to plumb the depths of any topic in any detail, but a book might provide the opportunity to explore more extensively. So that's what we've tried to do. In addition to the most popular topics, we've also included some subjects which, while not necessarily frequently mentioned by listeners, are some of my own favourites. After twenty years as a wildlife ranger with the Queensland Parks and Wildlife Service, I left in August 2003 to "pursue my own interests". Have a look at my website at www.drivingyouwild.com.au. I am greatly indebted to Steve Parish and his wonderful staff for giving me the opportunity to see one of those "interests" materialise as *Talking Wildlife*.

So if there's no homework – let's take some calls…

The oft-mentioned female Golden Orb-weaver.

For my father Owen

who, almost 50 years ago, showed me the little holes that bandicoots had dug in his "lawn".

While Australia laments the number of species that have been lost to the permanency of extinction within the last hundred years or so, there is, on the other side of the coin, a number of native species that are much less sensitive to huge changes and have survived in the built environment. There are even some that do better there, not because they have "adapted" but because the essential elements of food, shelter and reproductive opportunity are all available and predation is relatively low. We often hear people talk of native species having "adapted" to the urban environment. I know what they mean but I've always thought the concept misleading and quite wrong. When a bird lands and perches on a horizontal powerline, even if the line sways from the impact, it is not much different from the bird landing and perching on a thin horizontal tree branch, which will also sway in response.

Genuine adaptation is much more serious. One stunning example of adaptation to the human world was the rise and fall, in the UK, of the melanistic form of the Peppered Moth, *Biston betularia,* in response to the rise and subsequent fall in the effects of industrial pollution. Before coal soot blackened the environment, the common form of the moth was light in colour. By the late 1800s, the most common form of the moth in the cities was very dark. Birds picked off the light coloured moth from the blackened tree trunks while the then rare dark form was overlooked. The dark form prospered and became the dominant form. Following pollution reduction and consequent cleaning up of the cities, the commonest form of the moth again became the light form as the process reversed. There is good evidence that this change was copybook Darwinism and that the effects of bird predation had favoured the better camouflaged form.

But when a possum finds a suitable gap in a damaged or incomplete building and then regularly uses the internal cavity as a day shelter, it hasn't adapted, it's just sheltering in a safe spot. As possums are predominantly folivores (leaf eaters) and there is often a good diversity of edible plants anywhere where there are houses – why not shelter in the roof? You take your chances with predation anywhere you live and the urban environment is probably no worse or better than the non-urban.

The diversity of native species that are prepared to be our house guests, invited or otherwise, is quite remarkable. Most will cause at least some people to view them as a problem. In House Guests we look at the most commonly reported species and the best way of managing the event.

The Inside Jobs

Possums in the roof

Possums are often found sheltering in roofs and by now this should be the most successful domestic wildlife management issue in Australia. But sadly it isn't. It must always be remembered that a sound roof on a well designed, well maintained house will not allow access to possums. When they do get in, it is the building which is at fault not the wildlife. As houses and other buildings age, the effects of weathering such as storms and high winds can loosen timber and lift roofing iron and gaps can form as a result of termite damage or dry rot – and that's where the brushtail possum comes in. And out. It is the only marsupial known that provides an all-year-round, totally free of charge, building inspection service.

One of our own resident brushtail possums.

Q. *I hear noises in my roof and I think I've got possums up there. How do I get them out?*

A. Easily. But first, let's make sure they are possums.

Brushtail possums have a heavy thumping footfall as they cross the ceiling. This is because their pectoral and pelvic girdles are designed to operate most efficiently up in a tree. On the flat they look a bit like a toddler in a nappy. The legs swing out widely and the feet come down heavily. Up the tree it's all agility and grace. If there is more than one there will usually be some vocalisation as well. Sometimes described as "heavy breathing", the noises brushtails make sound like a long drawn out hollow hiss as well as staccato coughing. These are warning signals to other possums. When the need to warn becomes urgent the hiss has a screaming sound to it.

Q. *There are yellowish brown stains down the tongue-and-groove walls. Is that from the possums?*

A. Yes. Normally, possums do not urinate in or near the den site. When this happens it is the most extreme form of warning to another possum.

Generally, what happens is: the roof space has been occupied by a female; she has produced young; the young possum has reached maturity and is no longer recognised as the baby; that offspring is no longer welcome in mum's camp.

First the mother hisses at it to move out. The hiss can reach that thin screaming sound on occasions. Bub doesn't want to leave. (It's rather nice here except for the hissing!)

All the while, mum is scenting the entry point and surrounds with her personal Eau de Possum Parfume which she hopes will give the youngster the message. It doesn't. So eventually she resorts to Eau de Nuclear Parfume! and scents the area with urine which also includes powerful potions from anal scent glands.

This is the point we've reached when the streams flow down the wall.

Q. *I realise I have to fix the hole they're using to get in, but how do I get them out?*

A. You don't have to. They go out themselves – every night – unless it's absolutely teeming with rain.

When you have identified the access point (or points) and have prepared the appropriate material to block it off securely, you wait until the possums go foraging. It is their habit to go out each night to feed. They do shiftwork. Starting time varies with the quality of the territory, but it is safe to say that all self-respecting brushtail possums are out and about within the first hour of darkness.

Then you shinny up the ladder and do it. It need only be temporary but it needs to be firm. Then the permanent work can be done in daylight. I've used a 3 ply panel screwed over the offending gap.

The area needs to be washed with a good scent killer. Household bleach and a strong disinfectant mixed in equal proportions will work well. If this is not done, the evicted possum will try very noisily to re-enter the blocked doorway. They're not silly. It's just that their nose will be telling them "this is where I live" despite the fact that the nose won't go through the locked door. So they scratch frantically at it to see if it will open.

Q. *We heard that the possums only got into the roof because all their natural hollows in trees have been removed. Is that right?*

A. No. This sounds rather good, as if it's a form of Nature's revenge, but it's not so. The possums got into the roof simply because they could.

Studies by Dr Geoff Smith of the Queensland Department of Natural Resources have shown that hollow-nesting mammals never have only one den site. Brushtails can have up to five different daytime shelter sites. And that makes sense: it's insurance against loss. No matter how many hollows you have in the back yard, the hole in the roof will be investigated and used.

Q. *We were told that we could persuade the possum to leave by putting mothballs or camphor up in the roof. Does this work?*

A. No. I have no idea how this one originated but it doesn't work. The roof space is huge, so how many mothballs would you need?

Q. *Someone said a light up in the roof would keep the possum out. Does this work?*

A. No. A light left burning in the roof cavity will not repel anything, but it is likely to attract the attention of law enforcement people who are accustomed to seeing unlawful plants being grown under light in roofs. Apparently.

Q. *What about these electronic devices that are supposed to repel pests. Do they work?*

A. I really don't know, but I strongly suspect not.

There has been a flurry of excitement surrounding the use of ultrasound to repel pests, but the theory seems very flawed to me. Apart from the well known and famous examples, I haven't seen anything that demonstrates that native animals have "ultrasonic" hearing. Investigations have been carried out on dolphins and laboratory rats as well as a fairly large number of wild species overseas, but little or nothing on Australian wildlife, except microbats and perhaps flying-foxes. The evidence in the case of flying-foxes is that their auditory range is about the same as humans, so any noise you are going to use to try to scare them off will be able to be heard by humans, and that's a happy thought! Little bats definitely have hearing ranges well above that of humans and since laboratory rats have been measured as having a range from 5 Hz to 50 kHz (which is a huge range) it seems logical that rodents may have "ultrasonic" hearing.

That said, no one has satisfactorily explained why ultrasonic noise is more effective than any other kind of noise in annoying possums, rats or anything else. We know that possums can hear what we can hear, otherwise their own vocalisations, which we can hear, would be useless to them. So why not treat them to a loud sound we can hear? You'd think a fairly blunt electric plane used on very old dry hardwood would be a reasonable test, right? It has absolutely no effect. I once carried out this experiment quite unintentionally. I was wearing ear-muffs, but the noise was so horrendous I had to stop. The possums in the veranda roof were totally unfazed. Besides, the electronic pest deterrents that emit high frequency sounds are far more expensive than fixing the roof, even if you have it all done for you! All they are likely to do is drive the dog nuts because we know that dogs definitely hear up the top end.

Frog in the loo

Q. *I have a green frog in the toilet and although we've taken it outside on a number of occasions it keeps coming back. How can I persuade it to live outside?*

A. You can't. Persuasion isn't an option. Either it physically can't get into the house or it can.

For frogs, the loo is a great place despite what we use it for. Moisture and high humidity are the clues and cues. Aside from the usual needs of food, shelter and reproductive opportunity, frog survival depends on being able to maintain satisfactory hydration, or body water content. Frogs are very susceptible to dehydration. Their skin is a sieve so far as water is concerned. When a frog has lost about 40% of its body mass, it has lost most of its water and is doomed.

Staying close to water, and even in it if conditions are dry, is a guarantee of survival and survival is a powerful driving force. A fully screened house reduces the chances of the loo being occupied by frogs.

Frog in the loo.

Q. *Our house is fully screened, and we can't work out how the frog gets in. No matter how many times we put it out, it manages to return. How is it getting in?*

A. It's probably getting back via the toilet's ventilation pipe.

As unlikely as it sounds, Green Tree-frogs can squeeze through the narrowest of apertures. The pipe which protrudes above the roof from the loo and which is usually fitted on top with a little spherical wire hat is the culprit. Even though the gaps in the wire are only about 25 mm, this is a breeze for even the largest green frog. If you cover the top of the pipe with some finer mesh then take the frog outside, all should be well. Choose the leafiest part of the garden and, to make life secure for the frog, set up a bucket full of large coarse rocks and water for it to hide in. Keep a few inches of water in the bucket and check the water regularly. There isn't any way to make sure that the new home will be successful, but the more damp hidey-holes you provide in the garden the more likely at least one will fit the frog.

Q. *Is it really the same frog that comes back or is it another frog altogether?*

A. It could well be the same frog.

Without marking the frog in some way, it is not possible to be certain, but there seems to be good evidence of a homing ability in frogs, especially Green Tree-frogs. One explanation for this is that, like the aquatic chelonians (turtles), frogs may have chemo-receptor cells which "taste" the environment around them and send a message to the brain that they are either at home or not at home. If the message is "not at home", the frog keeps moving until the "at home" message is received. Experiments on frogs overseas have shown that even quite small species of frogs have returned home after being displaced by up to one kilometre. For a 50 mm long frog, that's quite a hike!

I was with Dr Phil Bird of the Queensland Frog Society one morning in pouring rain when we observed seven pairs of Green Tree-frogs in mating embrace all sitting on a mown grass surface. Only a year before this place had been a natural depression which, when it filled with water during rain, was a well known breeding site of this species. Clearly something was telling the frogs to be there even though their pond had gone. We were accompanied by the local politician who had arranged to have the ditch filled in to prevent mosquito breeding and who, on seeing the plight of the frogs, moved immediately to have their ditch re-dug. It was. The frogs now happily breed there again.

Q. *Our house has frogs everywhere, behind the pictures – everywhere. What can I do?*

A. Screening is the only solution.

There are plenty of people who wouldn't mind this in the least. Frogs are very popular and a house full of frogs is a house devoid of cockroaches. But if there is no choice and the climate is such that the windows need to be open at night, then insect screening is the only long term solution.

Little bats in the wall

Few modern houses sport a belfry, but the phenomenon of a wall full of microbats is not all that unusual. Microbats, the small insectivorous bats that use echolocation to track prey on the wing, shelter during the day in any suitably sized crevices, tree hollows and caves. Cavity walls of houses are ideal if there is a gap sufficient to allow their comings and goings.

Alan and Stacey Franks

Gould's Long-eared Bat.

Q. *There's something in the wall of our house. We hear faint skittering sounds and it sounds as if there are hundreds of them. What could they be?*

A. Insectivorous bats or microbats.

These are the tiny bats of the night that do an enormous service to humans by eating insects. They are very different from the giant bats or flying-foxes which roost in large conspicuous camps in trees. The little bats are rather flattened creatures and, anywhere the head will fit, the rest will too. In any given region there are usually a few species that are happy to colonise a cavity wall in a house if they can get in.

Q. *Will they do any harm in there?*

A. Yes, unfortunately they can.

As much as I love them, a large colony of microbats in the wall can eventually lead to serious problems. The bats fly out after dusk and immediately begin feeding. To rest, they return to the roost site in the wall and that's where the fun begins. As their food is digested and becomes waste it has to be excreted and this they do in the wall. While this doesn't sound great to us, it is all part of their grand plan as the scent of the latrine section at the bottom of the wall is an aid to their relocating the campsite.

If the species is one of the larger ones that form big roosting groups of five or six hundred animals, the latrine soon grows. There is nothing unhealthy for them in this, but there can be for us. Dried bat droppings can form an aerosol that, if inhaled, has disease problems for humans. The build up of guano can also do a great deal of damage to the wall, depending on the material it's constructed from.

Q. *Can we get a pest controller to fix the problem?*

A. A very experienced pest controller, yes, but I would check their method before engaging them.

Solving this problem to the satisfaction of all concerned needs special care and experience. The bats don't deserve to be harmed as they have only taken advantage of what seems to them to be an ideal home. They are also fully protected in all States of Australia which might raise the issue of offences and penalties if any are in fact harmed.

The walls can be freed of bats; the bats can be provided with an alternative roost site in the back yard; and they can go about their insect feeding work which is of great benefit to all of us. But it must be done properly.

Q. *Where do you start?*

A. You need to identify the flyout points.

A 7–8 mm gap is all that it takes to allow the passage of the small bats. These can occur as a result of the slight buckling of weatherboards over the years and can be seen from below, looking up. Sometimes the flyouts have been created by the design or execution of the building when a sloping gable roof fits over horizontal timber or fibre-cement weatherboards.

The best method of searching for the flyout points is to go outside and wait nearest the place where the bats have been heard. On dusk, the occupants will be seen emerging from the gaps and flitting off into the sky. The light will be quite dim by the time the exit is in full swing, but you will still be able to see where the bats are emerging.

Regardless of the nature of the gaps, they need to be filled. Modern gap-filling products are ideal and some work can be carried out in the daytime, but it is vital that the bats are able to escape. You need to leave an unfilled section wherever you have seen the bats emerging. The most used section is the best to leave untreated.

Q. *How do I know when all the bats are out?*

A. You don't and they won't be. This is the tricky bit.

Microbats, unlike most other nocturnal animals, don't go out and stay out all night. Some emerge, fly around and feed then return to the roost to rest while others take their turn in the sky. At any one time it has been estimated that as much as a third of the total colony will be in the roost site. So if the flyouts are blocked up even at night, about a third of the bats will be fatally trapped inside.

Q. *So how do I get them all out?*

A. With a bat-valve designed to allow the bats to emerge from the roost but prevent them from getting back in. This way all the bats will be outside the next day when the rest of the gap can be filled permanently.

A bat-valve can be made from a plastic garbage bag. Tape a section of the open end immediately above the flyout and tape the lower lip to the lower edge. Cut the bottom of the bag open to form a wide plastic tube. The bats will be able to slide down the tube from the taped opening and fly out, but when they come back, they will not be able to get back up into the roost. The bag should be trimmed to about a 600 mm tube.

The valve should be left in place for a few days and the bat roost monitored until you are sure none remain. If the familiar sounds are still being heard in the wall, it is certain that not all the flyout area has been correctly identified and you will have to look elsewhere for flyout points. When the wall is clear of bat sounds, remove the valve and fill the remaining escape hatch.

Q. *Can I do this at any time of the year?*

A. No. It is important to be certain that no dependent, non-flying young are in the roost.

The young of microbats are born in late spring and remain with their mothers until the end of January. For part of this time, the young can't fly. If roost sites are closed off at this time they are doomed. The best time to undertake the gentle bat eviction method is in autumn, or at least after February and before June when we are sure that all living bats are fully aerodynamic.

Q. *Where will they go if I evict them?*

A. We don't know, but we do care.

Most zoologists worry that at least some of the bats will perish if alternative roosting is not provided. This is not known for certain and it may be that most, if not all, will actually survive by simply finding little nooks and crannies in trees nearby. But, to be on the safe side, I would recommend that, if possible, an alternative bat-house be set up in a nearby tree. Several may be needed depending on the size of the camp.
I would recommend contacting Hollow Log Homes at Kenilworth, Queensland, **www.hollowloghomes.com.au**, for a bat-house design that really works.

Antechinuses in the drawers

People living in very bushy areas, usually close to a large and largely intact closed forest, sometimes have problems that most of us can only dream of. One of these is the "problem" of antechinuses. Antechinus is the genus of small (but not tiny) dasyurids which are carnivorous marsupials. "Antechinus" comes from Greek and means hedgehog-like.

When the first species was described to western science in 1842, the antechinuses were thought to belong to the mammal taxonomic order *Insectivora*, which included the European hedgehog, but they are quite distinctly separate. The extinct Tasmanian Tiger was the largest of the group in modern times and the Western Australian Numbat and Marsupial Mole are the most specialised.

Alan and Stacey Franks

Spot the kitchen Antechinus.

Female antechinuses nurse their young in nests made of plant material collected in hollows in a log or tree.

Cupboards and chests of drawers are great substitutes for tree and log hollows and will be used by mother antechinuses if the opportunity arises. Not everyone is happy about this.

Q. ***I've got these rat-like marsupials in my drawers. It looks like a mother and babies. How do I get rid of them.***

A. Very carefully, please!

Although some people could see them as nuisances, these animals are an incredibly precious part of the native fauna. Members of the group are generally very sensitive to "development" and are the first to go when natural habitats are changed. All males and most females live for about one year only.

I'd do nothing until all the nesting activity is over then I'd have the cupboard, chest of drawers, or whatever, antechinus-proofed. This means covering backs and bases with thin plywood glued and screwed into position to eliminate any gaps that might allow them entry. Modern gap-filling products should be used to fill wide spaces around pipes in stoves and walls, etc.

Prevention for next time is better than cure this time, especially after the animals have already set up house.

Alan and Stacey Franks of Hollow Log Homes have first-hand experience of antechinuses in their house and have also designed a nest-box specifically to provide the appropriate shelter for nursing mother antechinuses. I'd seek their advice.

Rats

Although some native rats will share our homes, the most common rat found in human dwellings by far is the introduced rat *Rattus rattus*. Its trademarks are the patter of little feet across ceilings and in walls. They also whistle and squeak and their presence is often accompanied by the dreaded sound of chewing. You hardly want to know what it is they're chewing, but it will be something you need. There are a number of proprietary products available from hardware and supermarket stores that can be used to control or eliminate rats from buildings and you may not have a lot of choices in the matter.

Some people prefer to use a catch-'em-alive trap and take the rodent for a drive. Unfortunately, this just transfers the problem to somewhere else or likely leads to a less than humane end for the translocated. Serious rat-proofing is the best method of management all round.

If there is any doubt about the identity of the rat, I strongly advise using a catch-'em-alive trap and taking the prisoner to the nearest expert for identification. Museums are usually happy to do this. If it turns out to be a protected native species, it can be released. For those without ready access to a museum or rodent expert, the *Field Guide to Australian Mammals* by Menkhorst and Knight (Oxford University Press) will help, but the decision on identification needs to be made with great care. Separating native species of *Rattus* from the introduced ones can be quite tricky.

Geckos

Fortunate are we whose houses are blessed with geckos. Over the years, I have had more calls from people who would like to get some geckos for their house than from those who want to see them gone. Certainly they will leave droppings clinging to the walls, but all the rest is winnings. You can pick a reptile dropping by the chalky tip. Reptiles do not have a separation of faeces and urine as mammals do. They concentrate their urine into solid white urea which looks very much like chalk but has a very pungent, acidy whiff. If the droppings tend to worry you, try to remember what they were before the gecko ate them!

Q. *We hear a loud metallic clicking in the house. Is it a gecko?*

A. **Yes, almost certainly so.**

The very loud metal sounding clicks are made by a friendly little lizard known to science as *Hemidactylus frenatus*, aka the Asian House Gecko. It was accidentally imported in cargo and probably household goods over a period of time in different Australian ports. It has been well known in Darwin and Townsville for many years and has spread rapidly in Queensland's south-east corner over the past 20 years.

The calls of the territorial males are really quite spectacular for a relatively small lizard. Fears have been held that it could move into non-urban areas and displace native geckos but this does not seem to be happening. Around Brisbane, the Sunshine Coast and Gold Coast, the House Gecko appears to be occupying only areas that had no house geckos. Where the native house gecko, *Gehyra dubia,* is common and abundant, as it is at my house, *Hemidactylus* appears not to venture. This may not last, of course, and there may come a time when conservation might demand control, but at present there is no cause for alarm. Australia has many other far more urgent issues than the Asian House Gecko. There was a time before geckos in my house when cockroaches ruled. As the geckos established a good population, the cockies were wiped out, and a good thing too. If you've got geckos you've got an asset.

Cockroaches don't stand much chance against the native house gecko.

Snakes

Walls and the spaces between the roof and ceiling are fabulous places for snakes. What bits of the world could be better than these to hunt (rats), hide (from predators) and shed the skin, which is a difficult time for snakes. Older houses often sport a roof area absolutely festooned with shed snake skins. Not all the owners of these houses know about this and perhaps that's not such a bad thing.

Paul Grimshaw

A particularly handsome rat catcher.

Q. *We hear a faint sort of hissing sound on the ceiling as if something is sliding across it. Could it be a snake?*

A. Yes, indeed. But it is really no cause for alarm. I know it's easy for me to say that, but it's not because I'm not frightened of snakes, it's because I know that with very few exceptions, the snake in the roof is of great benefit and of no great threat.

Australia has no mambas. They are South African species that you could actually meet up a mango tree. Their bites are very serious indeed – they are of about the same toxicity of our big browns (*Pseudonaja* species) – and will usually be fatal without proper treatment.

Fortunately, our dangerously venomous snakes don't climb all that well. Although there have been reports of Rough Scaled Snakes, *Tropidechis carinatus,* in up to 2 m high lantana, and although the members of *Hoplocephalus* – the Broad-headed Snake, Stephens Banded Snake and the Pale-headed Snake – can all climb well, the only snakes regularly recorded in suburban ceilings are the pythons and tree snakes.

All shed snake skins can be identified if you know what to look for and if you need one identified, your state museum is the best place to send it.

The real issues with snakes are the ones we have control over: sensible behaviour and proper first aid and treatment. The presence or absence of a snake is Nature's business.

Q. ***We have a snake in the lounge and we can't get anyone to come. What do we do?***

A. Isolate the snake.

Close all doors and windows in the room and pack towels under the doors. Once seen and alerted, the snake is most likely to try to hide somewhere. Now that you know which room the snake is in, and that it can't get out, you can calmly consider your options.

If there is a snake help service anywhere within a reasonable distance, wait for it. If that option is never going to happen, the best thing to do is to make up a snake hiding box. A cardboard carton of appropriate size, all taped shut, with a hole of appropriate size in one corner, is just the thing.

What you are going to do is convince the snake that it wants to slither through the hole in the corner of your box and hide. Find a broom, and go and put your gum boots on. If the snake is dangerously venomous, it will be a ground-hugging snake, but likely of some agility and speed. Nevertheless, if things go a bit wrong and it has a whack at you, you have absolute protection almost up to your knees.

Armed with your trusty broom and the box, enter the room and close the door behind you. Replace the towel under the door. Place the cardboard hide box against one wall, making sure that the hole designed to attract the snake is hard against the skirting board and that the box is stable. Some books on the top will keep it in place.

If the snake is visible, you will need to usher it toward the box. Unless it really feels cornered, it will tend to want to go away from you. It will tend to want to hide. If you feel it's going to be able to do this easily and in lots of different places, you may need to empty the room by passing its contents out the window to an assistant. If the snake gets a bit flighty, back right off and keep your distance. It can only be a problem if it bites you, and to do that it has to be close to you.

If you move fairly slowly and don't corner the beast, you will eventually have it in a position where it sees the opportunity to get out of harm's way (you) and into that safe box. Clamp a book over the hole. With the book firmly in place, turn the box so the hole is on top and march the lot outside and down the paddock. Return home with the book and next day the box will be empty. Don't take it far away: that achieves nothing save to drastically reduce the snake's chances of survival. And since you've gone to an awful lot of trouble to respect its right to life, why toss all your excellent work away? If your courage fails you, give the whole thing away and wait for help regardless of how long that help might take to arrive.

By chimineys!

Why chimneys aren't fitted with a fine mesh exclusion fence as a factory standard still amazes me. Certainly, there are thousands of chimneys Australia wide that have served solely as a chimney, providing a one-way passage for unwanted smoke, and nary a problem has arisen. But there are also thousands that have provided a source of serious concern for their owners, and some on more than one occasion.

A no-mesh invitation to a hollow-nester.

Once upon a time in south-east Queensland, there lived a very happy couple who decided to go on a three week holiday leaving their lovely house in the care of the local wildlife. This was a fine house. In its cream shag-pile carpeted living room was a magnificent stone fireplace with beautifully crafted stone chimney. The fireplace had provided romantic fires throughout three winters and the chimney was nicely coated inside from many a log being consumed by the flickering flames.

During their absence, a very large male brushtail possum mistook the chimney for a hollow tree and investigated same as a possible daytime shelter. Presumably to the tune of "Slip Sliding Away", it cascaded into the living room with a fair slice of sooty deposits and, alarmed at this apparent poor treatment, attempted to leave the house via every closed door and shut window, using every available piece of furniture and fittings.

Lace curtains failed as vines, but the artistic patterns of soot and droppings on the formerly cream carpet would have got runner-up as a Pro Hart television commercial.

Fortunately for Mr Brushtail, his misadventure occurred only a few days before the owners' homecoming, so he was still alive when I was called to take him "outdoors please".

Some chimney structures are such that brushtail possums can actually shelter in the "hollow" without falling all the way down. Possums have discovered this, and some people have discovered possums that have discovered this. If you are a person who discovers a possum that has discovered this, please do not attempt to "smoke it out". In my opinion, this is a technique invented in b-grade Hollywood westerns.

I once (only once thankfully) attended a rescue of a possum that was the victim of an attempted smoking out and, although it was alive, it took weeks for the damage to heal before he was eventually released to the wild. It doesn't work.

The list of accidental intruders via chimneys is fairly impressive. It includes Carpet Pythons, Green and Brown Tree Snakes, cockatoos, corellas, possums, gliders and kookaburras. Any species which shelters or nests in hollows is a potential house guest. You may live your entire life without unwanted wild guests arriving down an unguarded chimney , but you may not, as well, and prevention is always better than cure.

On the Outer

Windows of opportunity

So far as native birds are concerned, reflective plate-glass windows are a relatively new phenomenon, especially in Australia. Our windows can be responsible for two quite separate and very different problems and Wildlife Talkback regularly receives calls about both. Since the problems relate to very different aspects of birds' lives, the solution for one conflict will never work on the other.

Q. *We have a peewee that attacks our windows and it's driving us mad. How do you fix it?*

A. With some difficulty, I'm afraid.

It isn't just peewees (or magpie-larks or mudlarks) that will do this, but the peewee is definitely the bird most often reported on Wildlife Talkback. Other species include crows, fairy-wrens, catbirds, butcherbirds, Black-faced Cuckoo-shrike, Grey Shrike-thrush, Willy Wagtail, Bush Stone-curlew (once), a Brahminy Kite (as unlikely as that might sound) and I'm certain that, if records were taken for long enough, the list would continue to grow.

They are deliberately sparring with a reflection of themselves, which these birds are convinced is a rival. Their confidence is high and their determination to drive away the intruder make them pretty difficult to deter. Scaring them is totally out of the question. They're not scared of anything. Although the cause and effect seem obvious, there is still the puzzle of how precisely this behaviour is triggered.

Heck Lindsay

Not afraid of a fight!

Were it simply a case of "bird sees reflection, mistakes it for rival – beats hell out of rival", then just about every window in the country would be being belted. There is no shortage of peewees and their distribution includes most of Australia. Unless the phenomenon is random (unlikely, but possible), it seems that a sequence of events under the right circumstances leads to this behaviour. The window will be within the bird's defended territory area, and I think it likely that it discovers its reflection during a critical part of courtship, nest-building, egg-laying or hatching when there is a lot of reproductive investment and therefore the stakes are high.

Within a few days, the routine can become established and stopping it becomes harder and harder. Covering the windows usually doesn't work. I once covered all of the ground level windows of a building with two layers of brown paper and the bird simply tore at the paper until the window was again exposed. It was clear that this bird was no longer reacting to its reflection: it was finding its imaginary rival using visual and spatial cues such as the plants in the garden and other large stable structures to navigate to the conflict site.

Birds are well known for this. The best example is the one of flock-breeding birds, such as seagulls and terns, finding their own egg in an ocean of nest sites and eggs. It's not that they can pick their distinctive egg from all the others; it's that they've memorised all the visual clues in the area and, no matter where they land, they move around until the landscape patterns and objects all make sense and lead them to their egg.

For the peewee, the impulse to find that rival and beat it out of the area is very powerful. Whatever preventive method is used, it needs either to be more powerful (no naughty stuff now) or to confuse the bird as to exactly where the rival was. This can involve a bit of work but the obvious choice is to eliminate the original reflection, provide a brand new one in a different spot where it doesn't matter, rearranging everything to make the local landscape unfamiliar and very confusing. It has to be done at night when the bird is off duty. If the bird sees your work as it progresses, it simply adjusts its memory and will still find the rival. You won't have confused it one bit.

If possible, move all the large objects (outdoor furniture, etc.) around, placing them in different positions and angles to the house. Cover the window with opaque material (it won't be for too long) or festoon it with inflated toy balloons, and put out the largest mirror you can afford somewhere down the far end of the garden for the bird to find next day. Leave the new arrangements in place as long as necessary. If the bird is still thwacking away at the original site, you've failed to fool it and need to start again. A good move is to advise your neighbours of the problem and the solution or they may think you've been at the cooking sherry.

Owl faces, cat faces, plastic snakes, etc., have all been tried and all have failed! If all else fails, remember, the bird will eventually stop whether you do anything or not. Often, I have been told of a remedy that worked. Being me, I would try it on the next available case without success. After a few episodes, I realised that the remedy appeared to work because it was being applied just as the bird was about to stop anyway. Meanwhile, to get some respite, you can always park the car on the driveway and let it play tunes on the hubcaps for a while.

A peewee shapes up to a potential opponent.

Q. *Not long ago a kookaburra began charging into our windows. Is this common?*

A. It's not common but it is very well known to wildlife managers.

Again, the precise mechanism which triggers the behaviour is unknown. My own experience suggests that it occurs in young, non-breeding males that may perceive their reflection as a challenger. As many windows are screened and the screens get holed in the process, for a long time I thought it may have been nest-making behaviour; that perhaps the screen resembled the surface of a termite mound to them. Now I lean more to the young male theory.

Q. *It hits the window at a ferocious speed. Is it likely to hurt itself?*

A. It could. I have seen the remains of a window that a kookaburra crashed through. Both were damaged beyond repair.

Kingfishers, and that includes our two giants, the Laughing Kookaburra and the Blue-winged Kookaburra, are designed to take a fair bit of punishment at the cranial end.

Their nesting technique is embarrassingly macho. They will discover a termite mound, usually on a mature ironbark within the defended breeding territory and, taking a deep breath, charge it full throttle, hitting the almost concrete-hard surface to chip off enough of the surface to get a nesting hollow started. Once they've broken in they can then perch on the rim and dig out the rest of the brood chamber with that ice-pick that doubles as a beak. The smaller kingfishers, such as the Sacred and Forest have apparently been observed to suffer fatal injuries on rare occasions.

Q. *If it's so dangerous, why do they do it?*

A. Well, termite mounds are fabulously good incubation chambers.

The temperature inside is elevated by the termites and very stable, due to the nature of the mound's construction. It is insulation and central heating at its best. Have a look at one the next time you see a mound which has fallen from a tree and broken open. (This can happen because of age, and is sometimes caused by the termites getting fed up with constantly having to repair the chambers damaged by kingfisher and kookaburra excavations and seeking a quieter neighbourhood.) Few predators are able to raid the nest, given its height above ground and the difficulty of getting at it when being strafed by adults as you climb the tree. And if you get that far there is always the possibility that the entrance is blocked by the body of one of the adults wedged in the doorway.

Termite mound.

Q. *I'm afraid the bird will break the window and kill itself, so how can I stop it?*

A. The best way is to work out how easy it is to place a cushioning barrier between where the bird perches to begin the bombing run and the window.

If the circumstances permit, hang a wide sheet (old sheets, blankets, cheap hessian or heavy plastic sheeting) half a metre to a metre in front of the target window, so that if the bird is going to continue to charge the window, it will plop harmlessly into the soft barrier. Again, you may have to let your neighbours know what's happening or they might start to worry about your state of mind. I found that trying to rearrange the world around the window, as we did for the peewees, just didn't work all that well on the few kookaburras we tried it on, but it may be worthwhile taping a dozen inflated toy balloons to the target window at night, especially if the behaviour is in the very early stages.

In exceptional circumstances you may need to go to your State wildlife authority for formal permission to catch the bird and take it elsewhere, but I would exhaust all other strategies first as translocation can have a devastating effect for the bird and just catching and handling it has its risks.

Q. *We regularly have birds hit our large plate-glass window. Most are killed by the impact. Is there anything we can do?*

A. Yes! There certainly is.

This is by far the simplest of the window bashing cases and is very, very different from the last two. In these cases the birds are not attacking an imaginary foe in a territorial dispute; they are innocently going about their own business and mistake the reflected scene in the window for the real world. Instead of a clear flight path to the leafy hills in the distance they strike the reflection in very solid plate-glass.

In this case, a silhouette of an appropriate bird of prey will work to prevent bird strikes. All my tests have involved the use of a brown goshawk silhouette which I carefully copied to lifesize from illustrations in a bird book and then coloured black. The illustration to use is one of the bird flying overhead, not one of the bird perched in a tree. The latter is only likely to attract the attention of all the local honeyeaters who will spend half the day beating it up. Fix the overhead goshawk silhouette to the inside of the window and the tragedies cease. The remedy works because the birds are simply travelling and will immediately swerve to avoid the danger the silhouette represents.

A goshawk silhouette can be effective in discouraging birds from flying into windows.

Bird nests

Several of the introduced species of birds can be defined as "pests" because of their winner-takes-all behaviour. They recognise houses and other buildings as ideal places to build their nests. Architects unfamiliar with the habits of these birds are their greatest allies. Pigeons, sparrows, mynas and starlings all do it and all can be prevented with a bit of housework. Bird-proofing buildings is probably the single most useful tool to reduce the populations of true pest birds in a sustainable and humane way. Prevention of reproduction is better than cure of presence. If your house has lovely nooks, crannies and ledges tailor-made for pigeons, starlings, sparrows, and their ilk, my advice is to get busy with fine mesh bird wire and exclude them.

Alan and Stacey Franks

Welcome Swallow nest.

Ledges can be made nest- or roosting-proof by fitting a metal sheet at an angle of 45° between the vertical and horizontal surfaces, then painting it the same colour as the vertical surface. From the ground it's almost invisible. It works. Ever seen a pigeon walking at an angle of 45°? However, some of my feathered friends avail themselves of the built environment to achieve their ends just as the introduced pests do.

The two most talked about are the Welcome Swallow and the Fairy Martin. The swallow generally picks a nice narrow ledge and positions the nest so that the chicks can conveniently aim their rear ends over the side. The subsequent drizzle of droppings is the chief complaint. A temporary drip tray can be fitted under the nest to be removed when nesting is over.

The Fairy Martin is a slightly different case. To begin with they don't nest in a discreet pair, they jamboree! It's a huge convention and they have pottery classes. They glue their bottle-shaped nests of mud to the wall next to all their friends. Hundreds of pairs can gather together for the big show if the mud and food supply seems sufficient.

Flying things that bite us or eat our plants form a large part of their diet. To my mind, Fairy Martin and Welcome Swallow droppings are a lot safer in that form than in the form they took before the birds ate them. But not everyone is of that mind.

People who run commercial businesses such as resorts can get a bit twisted about it. OK! Bring out the clear plastic sheeting. It is very hard for a little martin to fix the house foundations to a flimsy plastic sheet that floats in the breeze, and when the birds have all decided they aren't wanted and flit off to help people somewhere else, the sheeting can be taken down, folded up and stored in case there is a next time.

Rude roostings

Q. *We've got birds that perch on the veranda rail and up in the rafters and their droppings are everywhere. How do we get rid of them?*

A. Relatively easily.

A single strand of fishing line, positioned at exactly the right height above the veranda rail will prevent the bird from perching there.

Q. *What's the right height?*

A. The right height is wherever the bird's waistline would be, so it depends on the species.

If the targeted species is a Feral Pigeon, the height needs to be about 100 mm. For a sparrow, it's half that. If it's a very wide veranda rail, two strands of line may be necessary. While many people would be delighted if there were a whole range of soft options, scary devices and electronic whatnots to convince some of my feathered friends to go about their business somewhere else, the simple fact is that prevention is far more effective than cure. Barriers work, so let's work on barriers. Depending on the nature of the problem, the angled surface is always useful. Birds find it very difficult to make use of a surface that's at an angle of 45°.

Fishing line stretched across the verandah railing will prevent birds from perching.

Metal spikes are a more expensive option.

What a beak, what a cheek, what a chew!

I remember the first time I saw the effects of the large, powerful beaks of the Sulphur-crested Cockatoos that had torn great chunks out of some new, very expensive timber doors. The doors' owners had described the birds' behaviour as "wanton destruction" and "sheer bloody vandalism" and I could sympathise. Most of the doors had to be replaced. The cockatoos had gouged out great slivers of timber from wherever they could reach. It was a heart-breaking scene. The owners had not long completed their dream home on acreage and while they were at work a few of my friends had paid them a visit.

Sulphur-crested Cockatoo.

I would be very cautious about using Western Red Cedar and similar timbers, even though they are light, strong and durable, for external French doors, having seen what short work the cockatoos can make of them. There is only one answer to this problem and that is to use seasoned hardwood rather than a timber that you can sink your thumbnail into. But it was the notion of "destruction" and "vandalism" that I spent a long time thinking about.

Free-ranging wild animals don't have the time or energy to engage in vandalism or wanton destruction. (Caged animals are different and can behave in very strange ways.)

For years I thought the chewing might have something to do with nesting. As it turns out, I am now certain that it does and the views expressed here have been supported by some careful observations by my friends Alan and Stacey Franks of Hollow Log Homes.

Some time back there was a bit of a public protest about the intended removal of some really magnificent Queensland Blue Gums *(Eucalyptus tereticornis)* near where I live.

The removal was allegedly necessary for safety reasons and duty of care. After some outrage followed by discussion and compromise, it was decided that only some large, threatening limbs of the trees needed to be removed, and peace returned to the streets. As the incident had created a bit of interest we used to comment on the trees and the change of heart each time we drove past them. As a result, we noticed that several members of the Psittaciformes (parrots and parrot-like birds) were paying a great deal of attention to the limb stumps.

Rainbow Lorikeets and Galahs were seen perched on the outer edge of the sawn stumps chewing away at the centre of what used to be the limb. It looked to us as if the formation of hollows isn't just left to Nature, but is speeded up considerably by the birds most likely to benefit. They are also those best equipped to help the work along. Those beaks aren't useful only for getting at food. So if chewing at the exposed end of cut or snapped off branches can speed up the hollow-forming process, it might just make sense to try it on any timber, including doors! Few activities of wild animals have no purpose.

If your windows and doors are already up and the cockies start chewing, you will need to immediately erect barriers to protect the timber. I long ago gave up on smearing timber with hot chilli and mustard and the like as it seemed to fail too often to be worthwhile. Another strategy is to ask builders whether they are aware that the chosen material may be unsuitable for the purpose.

Chewing for the future.

Things that go bump

Q*. We don't have possums in the roof. We have them on the roof. They wake us up every morning thundering over the house. Is there a simple way to prevent this?*

A. Yes, it's usually quite easy.

If the access to the roof is from overhanging branches of trees, you will need to trim these back two metres from the edge of the roof. Brushtails won't attempt a two-metre leap. If the access is via the powerlines, first check with your power supplier to be sure that any remedy is lawful and safe. A length of 10 mm polypipe, slitted and slipped over the line, will rotate in a very unstable fashion if the possum attempts to clamber along it. It won't cause the possum problems as their fifth limb (the prehensile tail) is always feeling cautiously behind should they find their feet are holding onto nothing but fresh air. If this happens the tail automatically grabs the line behind, they climb up it and reverse their way back along the wire. If the polypipe is installed during the daytime, the possum won't be able to get out that way so won't be coming home that way. Remember, it is unlawful and dangerous to fit anything to powerlines. Always seek help from your power supplier.

Unlawful and ineffective.

This one (demonstrated using some rope) works.

The Back Yard

Every time I venture to the local shops, I pass a back yard that absolutely amazes me. It is mown from one end to the other, right up to the three perimeter fences and beyond. Except for the solitary Hills Hoist, nothing grows above five centimetres high. As I write, we've had some good late summer rain, so it is all green. When El Niño returns it will be all brown. Around the old Queenslander is a low fluffy frill of geraniums. Nothing else. Some years ago when cat bashing had reached the levels of lunacy, I showed a slide of this yard as part of a talk on wildlife conservation in suburbia and awarded the yard the Most Wildlife Friendly Back Yard because I'd never seen a cat in it. In the 13 years I've looked at it as I drive past, it is usually completely empty, but on rare occasions I've seen the odd peewee, magpie, Willy Wagtail, House Sparrow and Spotted Turtle-Dove. That's not too bad for a habitat of two plant species.

I really couldn't live like that. For me the back yard is where it all begins. After all it's the part of Nature you see the most often and for the longest stretches. I know they get a lot of great press and everyone's tired of their getting so much attention, but I still thrill at the sight of a Koala up our Grey Gum first thing in the morning.

No cats – no nothing.

I took a bit of time away from these scribblings last week to get some photographs of the caterpillar of the Australian Leafwing Butterfly, its chrysalis and the adult as it emerged. There are few butterfly caterpillars that can match the stunning beauty of the leafwings. Some of the original type material of a sawfly that was only scientifically discovered and described in 1999 came from my back yard. Two-thirds of the spider species in my back yard are yet to be described and there's a lovely flightless cricket up the back somewhere that needs describing too.

This is not acreage; it is, according to the title deed, 1421 square metres or 56 perches in the old money. A bit more than an average quarter-acre. But we don't own a mower and if any plant that comes up is locally indigenous, it lives its life out here. I'm presently trying to catalogue the moths of our back yard. It will take me the rest of my life, there are so many species and so few moth-ers. The back yard for me is the centre of my natural universe. For some people it is territory to be fought over with the wildlife that wants to share it with them. Our policy is, if it's a local native it has right of way. Not everyone feels that way....

Dux of the pool

Why you can have an in-ground swimming pool for twenty years or so and then wake up one morning to find that it has been located by a group of Australian Wood Ducks is one of life's great mysteries.

You will notice too, that they are perfectly healthy wood ducks with all their faculties and functions intact.

Duck poooool.

I used to like to think they were refugees from the duck hunting season, tired of being shot at and desirous of some peace and quiet. And on finding the pool they then set about expressing their robust opinion about duck hunting and what it's like to be shot at.

Like brilliant comedian John Clarke, I also like to think that duck hunting is about shooting at ducks rather than actually shooting the "...beautifully winged quacking persons".

Sometimes, the ducks find the pool much earlier in its lifespan than twenty years, but the results are the same. After a few days you can't swim in the pool, unless you want to go through the motions, and the pool filter system is struggling with the extra load.

The common approach is to see the ducks as the problem. Well, they are, but if you want a happy outcome, you need to know what is compelling them to be there. And "compelling" is the right word.

Here is a list of things I have seen and heard that you can try. By all means read them, but please don't try them. There is a completely harmless and effective method and we will get to it shortly, but, meanwhile, here we go....

Get a dog.
Chase them yourself.
Join the duck hunters.
And, the daddy of them all, get a goshawk silhouette!

I looked in an awful lot of shops for a goshawk silhouette and guess what? However, for about $40.00, you can now actually buy such a device these days. One mob that sells them guarantees your money back without quibble when it's found not to work. The same website warned that it won't work at night because the ducks can't see it. Oh, yes they can, but this is very good marketing because all new dux of the pool cases I looked at had initially happened at night and most often on nights with a fair bit of moon.

If we take these scare tactics to their logical conclusion, there should be no ducks (or other "troublesome" wildlife) wherever there are birds of prey. Take a step further and you shouldn't have wildebeest anywhere where there are lions.

I made and rigged up a goshawk silhouette in a yard with a pool and half a dozen wood ducks and, after two days, they were camped in the shade of the thing. Initially they were a bit wary, of me as much as the silhouette, but they made no attempt to quit the place.

If you're going to get a dog, check with the RSPCA first and they will tell you all the right reasons for getting a dog. If all goes well, by all means get a dog but get one that can fly. That way the dog can keep chasing the ducks far enough for them to forget where your pool was. Otherwise they just do a few circuits till the dog wears out and then go back to where they were. In your pool. The dog will always give up first.

Way back in my early career, rather than tell people with problems to do all these recommended things, I would go to their homes on weekends and apply the remedies myself to observe the results. While I never actually took a dog, I did chase the ducks myself. I tried all of the remedies and found almost all to be as effective as if I'd tied garlic around the pool owner's neck. It was ditto for camphor, moth balls, hot English mustard and quassia spray (very bitter; made from bark chips from the quassia tree) too.

About this time (1985) I was reading about ethology and the works of Konrad Lorenz and Nikolaas Tinbergen, and decided that if any progress was going to be made in wildlife management, we needed some science or at least some careful observations. If the wood ducks were going to be seen off the pool premises we needed to know why they were there in the first place.

In the mid 1980s, studying wildlife in the urban environment was not considered to be real science or real zoology. Real zoology happened out in the bush. As a result no studies had been undertaken on any of the problems that faced me on a daily basis. All that is changing. As the bush shrinks (code for "land is cleared"), the urban environment is more and more being seen as a real place as well. I make regular reference to my friend Dr Darryl Jones of Griffith University, primarily because he was one of the first PhD recipients not only to recognise this but to be brave enough to establish the Suburban Wildlife Research Group at his university.

But back to the ducks. The swimming pools I visited at first just happened to be all surrounded by lush lawn and, as the birds were feeding on the grass, I got side-tracked for a while. I thought food. As fortune would have it, two problem pools in a row were set in tiled surrounds. No sign of any grass. So it was solely the pool.

Then, on a trip across the Brisbane River on the Moggill ferry, I caught sight of a pair of Australian Wood Ducks with about five half-grown ducklings, paddling along near the bank. Overhead went a Brahminy Kite and the penny dropped. While Brahminy Kites don't regularly prey on ducks as far as I know, their shape is unmistakeable and the ducks knew what to do. They dived!

I was then convinced that the pool or the water was "shelter" or safety in Wood Duck language. The three cardinal principles of life are food, shelter and reproductive opportunity. I then assumed that they located this form of shelter or safety by sight. Few bird species have any sense of smell and the taxonomic order Anseriformes (the ornithological term for waterfowl such as ducks, geese and swans) certainly don't, so the eyes must have it. And given that the visual signal for water is the way light reflects from its surface, I thought I should be able to move the ducks on if I could stop their eyes receiving the safe-shelter signal.

At the next available opportunity, I bought enough black plastic sheeting from the hardware shop to cover the pool and bingo! All gone within 15 minutes. Plastic sheeting to cover a pool costs about the same as the bird scarer hawk, but it works both day and night and the great advantage is that it can be used as a treatment, rather than a deterrent. For deterrents to work, the target species has to know what the consequences will be. This is the theory behind crime and punishment and since crime is unknown in Nature (except in the funny monkey – humans)... well, you can guess the rest.

There is another remedy for the dux of the pool problem that is very worthwhile considering should it arise at your castle. First ask yourself how many times in the past year or so the pool has been used for the purpose for which it was constructed. If the answer is disappointing, empty the water and fill the whole thing in with some good quality soil and grow some beaut organic vegies. Wood ducks especially love young lettuces.

Bird calls

Most native birds are diurnal, or active during the daytime. It is also true that more than half of Australia's vertebrate animals are birds. It is not surprising, therefore, that almost half of all calls to Wildlife Talkback are about birds. Of these, half also are requests for identification. This can be terrifying. Australia has recorded about 850 bird species and, to complicate things, most birds of a species are distinctively different to look at depending on whether they are adult or immature, male or female, breeding or non-breeding.

Then there are the calls or songs. No bird I know only has one single song. Some have a dozen and some calls of these species are heard only very rarely. Many species, far more than you might imagine, also do impersonations of other birds.

Steve Parish

Pied Butcherbird – singers of renown.

The pre-dawn singer

Over the past 12 years, the bird which has been recorded most by listeners and played to me over the air is definitely the Pied Butcherbird.

I think the main reason for this is that, during the early part of the nesting season, the Pied Butcherbird begins its early morning song well before sun-up and it often changes its early morning song during this time. It still sounds like a Pied Butcherbird but it ends up singing a melody which is really quite different from the original and not at all like the advertising call that ends with the characteristic "sweet dee dee!".

Often, people are trying to work out songs from the description in the bird books. Australia is fortunate to have some very good field guides. All attempt to give some sort of description of calls and songs. This is incredibly difficult. Here's an example. To me the most common advertising call of the Pied Butcherbird in south-east Queensland sounds like "Dee – dah – doh – dum dum…. Sweet dee dee!" The reader has no idea what pitch each word is being sung at, or the length of each note or the intervals in between. It would be a lot more accurate to represent the same Pied Butcherbird call like this…

But then not everyone reads music.

Most calls from listeners are based on descriptions and sometimes on a bird that was seen days, weeks, months or even years ago. This makes things difficult. There's the problem of human memory. Nothing terrifies me more than the opening line, "I saw this bird when I was on holidays and it's not in my bird book". The chances of a bird being seen anywhere in Australia and simultaneously not appearing in the current edition of the field guides by Morcombe, Slater, Simpson and Day or Pizzy and Knight are very remote. It does happen, but only on very, very rare occasions. There are just so many excellent bird observers out there that when a genuine first record of a "new" bird occurs, the flurry of excitement is heard far and wide and always at greater than the speed of light.

Flights of fancy... hummingbirds

Passionate beliefs are hard to shake. During the past ten years, a number of callers have insisted on their having seen a hummingbird. Sightings have ranged from suburban Brisbane to far north Queensland. There are three explanations. The first is that the person has actually seen a hummingbird; the second is that they have seen a small native bird that hovers like a hummingbird; and the third is that they have seen a hummingbird hawkmoth.

In the second and third possibilities, the view of the bird or moth is fleeting. No one can be expected to get a detailed description and over a period of time, the sighting becomes more and more like a hummingbird in the mind of the beholder. Native birds which could easily be mistaken for a hummingbird by the casual observer include the Eastern Spinebill, a colourful bird with a long, curved beak which does hover over flowers and the Yellow-bellied Sunbird which is very hummingbird-like in behaviour.

Several species of hawkmoth are large moths that behave just like hummingbirds. My own first sighting of one caused me great excitement. My second sighting caused me to catch the moth, cool it in the freezer, photograph it for identification and let it go. The photo was identified at the Queensland Museum as that of *Macroglossum micacea*. There is no common name, but it could go by the name of Fluffy-tailed Hummingbird Hawkmoth. It's a bit cumbersome but nevertheless descriptive. The hummingbird hawkmoths of the genus *Cephonodes* are really quite startling. Like *Macroglossum* they fly in daylight, which is not what we expect of moths, and they have brightly coloured bodies and almost clear wings. *Cephonodes kingii* is bright yellow and black with rusty tipped clear wings and would get my vote as a hummingbird any day if I didn't know about the moth.

The first possibility and the hardest to shake is that the person has seen a hummingbird. The chances of a vagrant hummingbird accidentally reaching Australia alive are very remote and can be safely discounted. Sightings of vagrant hummingbirds are all confined to the New World. One suggestion made to me by a listener was the possibility that someone had smuggled a hummingbird into the country and let it go. I talked to some friends in the wildlife park industry about the difficulties of bringing a live hummingbird to Australia legally, let alone by smuggling, and logistically it would be very difficult. As far as I can tell it has never been contemplated, let alone attempted.

You saw what? Where?

Various well-known aviary birds are regularly recorded out of their normal distribution wherever there are houses and people. Some escapees have now bred into fairly large populations and there are those who choose to believe that these flocks have occurred by natural expansion. It is instructive to note that, in the case of natural range increase, the species tends to be recorded continuously between the original area of distribution and the new. The species which most frequently cause comment are Long-billed Corellas, Port Lincoln Parrots and Twenty Eight Parrots (both species of Australian Ringnecks) seen many hundreds of kilometres from where "the bird book says they should be".
In my own bird feeder in south-east Queensland, we have seen these as well as Major Mitchell's Cockatoo, Superb Parrot, Red-winged Parrot, Crimson Rosella, Eastern Rosella (all out of their normal range) and Indian Ringnecked Parrots and African Lovebirds (both out of their normal countries).

Steve Parish

Aviary escapees have formed very large populations far from their former natural range.

Recently there have been sightings of Gang Gang Cockatoos in south-east Queensland. Some observers would like to think they flew there unaided, but my wager is on the cager.

Vagrancies do occur and when they do they can be exciting and within a few hours the word is out.

A Franklin's Gull was spotted at South Bank in inner Brisbane in 1998 and I decided to go and have a look. It looked like our familiar Silver Gull but smaller with red tip to a black beak and (you couldn't miss this bit) a shiny black head with narrow white crescents above and below each eye. By the time I got there, South Bank was crawling with people wearing binoculars around their necks and it took a while to get to see the bird – we all had to say hello to each other and if it happened to be someone you hadn't seen much of, a whole lot of hand shaking needed to accompany the greetings.

What's all the fuss about?

Franklin's Gull, a native of the Americas, has been recorded a few times in Australia but, in October 1997, a Filipino bird, the Blue Rock Thrush, was seen at Noosa National Park on Queensland's Sunshine Coast. This was a first in recorded history.

Every day there is someone birding. There are a great many excellent bird observers and for this reason it is unlikely that a large number of unidentifiable birds are flitting about. But I keep looking just in case.

Wildlife Talkback has given me a reputation with many people as a skilled birder and a skilled imitator of birds. I'm not. My friend Paul Grimshaw is and I will be forever grateful for the long hours we have spent in the bush. Paul's imitation of a Spotted Quail-thrush is so authentic that on one occasion a male of the species mistook his calling for a female and tried to feed him. That is a skilled birder!

Back to the problems of memory. Very early in the piece, I had a call from a gentleman who described a tiny bird with a huge voice and a bright red head. Before I had a chance to jump to the conclusion that he had seen a Scarlet Honeyeater, he finished the description: bright red head, black wings and white underneath. That wasn't my idea of a Scarlet Honeyeater, but bright red head… tiny… what else? "I recorded it," he said. Fantastic! Let's play the recording. Out of the tape player came the unmistakeable call (scratchy and re-translated via his phone though the telephone lines and into my headphones, but unmistakeably a Mistletoebird!) "Ah!" I said. "You mean bright red throat." "No mate. I was looking at it. I know what I saw. Red head, not throat".

After a little while I gave up. But I know what happens. I've done it to myself.

In about April 1990, spurred by some vague nonsense I'd read somewhere I was in Birkdale looking for bitterns in the "primeval swamp" down Collingwood Road (it's all gone now) when I spotted a dragonfly I knew I'd never seen before.

This turned out to be the real value of the outing. It was perched on a twig and I had a good long look at it. I stored in my memory what I thought were the distinctive features of the insect. Body and head – black. Black triangle shape in hind wings with a distinctive horizontal gold bar through the black triangle. The fore wings had perfect round black dots in them very like the Big Greasy – a butterfly.

Later in the week I went through the dragonfly drawers at the Queensland Museum and to my delight I couldn't find a specimen of my dragonfly. I had to wait until November that year to see it again and this time I was ready. I'd become so fascinated by dragonflies that I had a catching net and a bagful of enthusiasm. I caught it, took it out carefully for a very close look and was a bit disappointed to see that what I remembered as perfect round black dots, like the Big Greasy, were nothing of the sort. They were ragged edged blotches and when I went back to the museum and checked the trays again there were about a dozen specimens of the Yellow Striped Flutterwing, *Rhyothemis phyllis*.

Yellow Striped Flutterwing, Rhyothemis phyllis.

Years later I heard a memory expert say something like "memory isn't what actually happened, it's what you stored back in the memory bank the last time you dragged it out for review" or something like that. Makes sense to me. And I'm also now convinced that some things that we think are memories are actually only what we want them to be.

Talkback bird calls have also been extremely enlightening and valuable. In 2003, we had records of juvenile Common Koels being fed by parasitised Blue-faced Honeyeaters in June – a phenomenon not previously recorded as far as I know. Not the bit about the Blue-faced Honeyeaters, the bit about June! Rarely have koels been recorded in the south-east corner much after April.

Leaving out calls and songs, the birds which have attracted the most requests to identify them from sight are Blue-faced Honeyeater, Crested Hawk (Pacific Baza), Channel-billed Cuckoo, Common Koel and Spangled Drongo in the first row, with Black-faced Cuckoo-shrike, Pheasant Coucal, Pied Currawong and the Indian Myna/Noisy Miner confusion close behind. Magpies, butcherbirds, Torresian Crows and Tawny Frogmouths are all popular, but apart from the error that frogmouths are owls, these birds seem well known and most calls about them are to relate or share experiences and observations about behaviour.

Steve Parish

Tawny Frogmouth.

The human is a funny monkey. Birds that are sighted intermittently in the back yard are talked of as having come and gone – it's the same with Koalas – and, even if there are Pale-headed Rosellas or Sacred Kingfishers in the bushland over the hill, provided they are not seen in our back yard, then "we don't see them any more – they're gone!"

The truly migratory bush birds certainly do this (see Cloud Cuckoo Land on page119).

Some regular Q and A

Q. *Where have all the sparrows gone?*

A. They've been drastically reduced by good pest management.

It's true! When I was a boy, the most common birds we knew were sparrows and pigeons. There are still large flocks of pigeons but the House Sparrow is a much reduced species in the bigger cities. It is now much easier to spot a sparrow on a competition twitch in smaller country towns. (Check out twitching at www.birdsqueensland.org.au/twitch.html on the internet.) In fact, on one competition bird twitch we forgot to keep an eye out in the rural areas and ended in Brisbane minus the sparrow on our list. How embarrassing!

I don't know that there are any data to support this, but it's been my experience that House Sparrows have declined markedly due to the efforts of pest companies employed to bird-proof the built environment. It makes sense. All birds have lice and the sparrow is no exception. Their lice also like to suck on humans if they get the chance so eliminating sparrow nests around the house also eliminates the lice. The design of some older colonial-style houses provided many nooks and crannies suitable for sparrow nests. Newer designs have tended to provide fewer nesting opportunities.

Interestingly, many years ago, I was told by a primary school teacher that cities paid bounties on dead sparrows so little boys would get their Daisy air rifles and pop them off for the money. The little boys were no fools. They hunted the sparrows **after** the peak breeding season. No sense in killing all the sparrows – they financed the school holidays. This is one of the major reasons why bounty systems fail.

House Sparrows were once in large numbers.

Q. *I've seen a bird in my back yard and I can't find it in my bird book.*

A. What bird book are you looking in?

It is really important to use a good quality field guide to identify a bird. There are lots of books on birds, but many are collections of photos of particularly attractive or very commonly seen birds. Your bird book may well simply not record the species you are looking at. It is also very useful to make a sketch of the bird. It doesn't matter whether you think you draw well or not.

It is more important that you put notes around the sketch. These are the essentials: *Overall size of bird.* Relate to well known birds, e.g. magpie-sized, sparrow-sized, or eagle-sized. *Length and shape of beak.* Give size of beak relative to head, e.g., straight beak same length as head; beak five times length of head; curved down. *Legs.* Long and thin, etc. *Tail length and shape*. Tail about half length of body, or somesuch.

Then draw the areas divided by colour and note the colours. Even if the sketch bears no real resemblance to the original bird, a keen birder will be able to work it out. I love these sketches. I have a collection from listeners and treasure them.

Q. *What's the black and white bird with the big yellow eyes, looks like a cross between a magpie and a crow?*

A. It's a Pied Currawong.

Q. *Why haven't I seen one before?*

A. I don't know.

Steve Parish

Pied Currawong.

Turkish delight

The Australian Brush Turkey, a relative of both the endangered Malleefowl of the southern arid country and the threatened Orange-footed Scrubfowl of the tropical forests of northern Australia, is one of the most impressive birds of suburban Queensland and northern New South Wales. For those unfamiliar with the male bird's industrious, gender-equity, nesting program, there's a real treat in store.

In their natural habitat of closed forest, both sexes spend the day raking through the leaf litter in search of invertebrates, snapping them up as the unfortunate wriggling things give themselves away, and supplementing their protein with any fruits and seeds that appear even vaguely edible. They are essentially solitary in nature until sundown when they tend to form orderly queues and amble over to the current roosting tree and, by dint of flaps and jumps, ascend the tree to their pre-ordained perch for the night.

Giblets, our male Brush Turkey.

Obviously there is some sort of hierarchical system of roosting as is displayed in the proverbial chook house and this is not surprising as the serious ornithologists are convinced that the Australian Brush Turkey, *Alectura lathami,* also known as Scrub Turkey and less often as a Bush Turkey, is most closely related to the Galliformes which includes chickens and pheasants. I'm not. I don't think they're even really birds. But that aside, there is a definite advantage in being Top Turk in a tree full of roosting birds because it doesn't take too much imagination to work out what can happen to the birds who are forced to roost at the very bottom. It could be at the bottom of an awful lot of bottoms.

Himself is about domestic rooster size, an overall dullish blue-black with a reddish featherless face and neck and a yellow wattle at the point where the naked neck meets the feathered body.

The tail is a vertical fan and looks nothing like the south-end of any other north-bound Australian bird. Herself is similar but without the yellow wattle and her head and neck never go beyond a rusty pink however enthusiastic she gets about anything.

In breeding colours, his head and neck are bright red and the wattle a brilliant canary yellow and capable of distending almost to his knees. During frequent sprints to see off rivals, the dangling wattle can threaten to tangle the considerable feet and legs and it is truly a wonder of Nature that some don't regularly go base over apex in the mad tag games that are played at high speed.

They always run in preference to flying which they don't do at all well as adults, despite the fact that they fly superbly the day they hatch. That always raises a few eyebrows! They fly on their birthday? Day one? Yep! And they are all alone. Without doubt the most precocious of all Australian birds, the babies, which resemble quail rather than their parents, are plain brown and fully feathered and fly strongly from birth.

I once "rescued" one and, in the process of releasing it, was treated to the sight of this freshly hatched chick shooting almost vertically up and over a single storey building, and it just kept going. Its flight must have been more than 50 metres, a feat generally beyond the capability of the average adult. Now it had to somehow dodge the numerous perils of early life: dogs, cats, in-ground swimming pools, not to mention a fair range of natural predators as well.

The survivors are those that keep a very low profile, foraging in the early light of dawn and at dusk and hiding for the rest of the time in dense shrubbery. There is also undoubtedly an element of luck. The good news is that if they do survive this crucial two to three months, they are pretty much in the clear and only have to worry about crossing roads and pedantic gardeners.

If turkeys just raked a bit and roosted in the trees, it is unlikely that anyone would notice them all that much even in suburbia, but it is **his nesting** that brings them to the attention of the landlord.

These birds long ago abandoned the debilitating and risky practice of sitting on a clutch of eggs for three weeks at a time and opted for alternative methods of incubation. As Nature's nod to female liberation, the female concentrates on matters such as her grooming and keeping her figure to ensure she is ravishingly attractive to the opposite sex while the males of the megapodes accept all domestic responsibilities. There are no unattractive, undesirable female Brush Turkeys, a fact which has not escaped his attention, let me assure you.

Aeons ago, evolution alerted turkeys and their family to the fact that the complex activity of composting vegetable material produces heat.

How it does this is too big for this small book. Heat incubates bird eggs, and the rest, as they say, is natural history. Other birds can't understand why he spends so much time and energy raking half the countryside into one heap and he can't understand why they still sit on their eggs looking bored. Raking is not boring. It's cardio-vascular fitness at its best. And it's fun, if you don't own a carefully planned and expensive garden with a lot of Law and Order issues.

Megapodes aren't called megapodes for nothing – the word means "large feet". Although they work perfectly well as feet, they also double as rakes and very good rakes they are too. It only takes a few days of concentrated raking to accumulate a one tonne pile of glorious compost. A week's worth can equal three tonnes! These birds have a work ethic that makes ants look slovenly.

A Brush Turkey mound can be several tonnes.

Although nest mounding can be triggered almost any time of the year by rain or the discovery of suitable nesting material, raking activity outside the September to February Nesting Festival is often perfunctory and usually fizzles out. During the festival period, however, raking is serious business and a rakish Turk is not easily deterred. Many a human has fought the mound and lost. The reason is that those who take on the task of stopping mound building make the mistake of believing they are up against the bird. They are not. They are up against Nature. Male Brush Turkey brains are divided into four main sections: food, roost, rivals and sex, with sex being a very large section containing the detailed plans of how to rake a suburban garden into one single pile and attract every female turkey in the district. If you look very carefully at this plan, you will find that it is signed "Mum Nature".

Both male and female have heat sensor pits in the roof of their beaks. These pits detect temperature differences of only a fraction of a degree Celsius. They are pre-set to receive temperatures between about 28° and 33°.

Not my choice of plants, but definitely turkey-proof.

Above and below that causes dissatisfaction to the owner of the beak. In **his** case, if it is below 28° he will start madly raking fresh material to get the temperature up, or if it is over 33° he will pull the thing apart to cool it down. However, he rarely has to do anything, because his original raking usually consists of exactly the right mix of greenery, soil and moisture to produce the desired result. He regularly tests it, though, just to be sure. This is done by sticking the head (maybe that's why it's almost featherless) deep into the mound and scooping up a gobful. His sensor pits do the rest.

She also is programmed to check mound temperatures. After all, you wouldn't put your eggs in an unsatisfactory mound, would you? So she checks. It is the shape of the mound that attracts her attention, whereupon she digs a bit of a test hole and scoops up a beakful. What she doesn't notice is that he is hiding behind the great big tree on the perimeter of the mound. (Turkey mounds are always built in full shade to reduce drying and close to a big tree to increase spying.) When she spies the mound, he spies her.

As soon as she tests the mound, he's out from behind the big tree and it's case of "YOU'RE HOME EARLY!" Later she'll return to dig a hole and lay the egg and during this process he mates with her again. The next generation of Turks are on their way.

It is a tortuous trip though. Darryl Jones, who told me all this and whose PhD is on this very subject, was able to estimate that only about 1 in 200 eggs laid in a mound ever sees adulthood. This means that, statistically, every adult Brush Turkey you look at has had 199 failed siblings. To me, that makes each bird special. But for the gardener obsessed with law and order, these statistics and my fascination for the birds pale into insignificance as thousands of dollars worth of landscaped garden disappears over the road and under the neighbour's giant mango tree. Yes, over the road. More than once I've seen Turk raking across a busy road, the perfect nesting material (garden mulch) having been on one side and the ideal mound site on the other.

Q. *We are trying to establish a rainforest garden and this turkey just digs everything out. Is there any way to stop him?*

A. Yes. And lawfully as well.

Australian Brush Turkeys are protected wildlife in all States in which they occur. This does not mean that they can't be interfered with ever, it just means that they should not be interfered with unless circumstances are exceptional. As there are lots of useful strategies to create a happy working relationship between gardener and turkey, I would never consider his raking the suburban garden exceptional circumstances. Exceptional circumstances might be a farmer's total loss of crop and livelihood, but even in fruit-growing areas there are alternatives to harm.

Knowing as much as possible about his habits is very useful, and from there on it's a case of nailing everything down. A very useful modern landscape material is known in my part of the world as "oversized biscuit". It looks for all the world like large smooth river gravel but it is lumps of sandstone turned up during earth-moving and tumbled around in a concrete mixer to knock off all the corners. I prefer it because river gravel extraction can be a lot more environmentally damaging than this alternative. It comes in mixed sizes and will hold any mulch and plants in place against the most interested turkey.

Q. *We tried putting a sprinkler on his mound, but that didn't work. Why not?*

A. This is just unsound practice.

Putting a sprinkler on his mound will add moisture and have him rake harder to incorporate more material or cover the mound to prevent cooling. He won't abandon the mound, that's his boudoir and source of all joy! Putting barriers in the way of his raking can keep you amused but it rarely stops him.

If the mound has been active for many days or weeks, it is best to leave him to it and look to see what you can nail down, which involves the use of big heavy stones already mentioned and cheap wire mesh, the kind usually used for chicken runs. Place the large stones around the bases of the plants you don't want moved. Peg the wire mesh over the mulch. It's cheap and very effective. Turk will quickly discover that neither the rocks nor the mesh can be raked and give it away. The long term plan should be to create a garden that is turkey-stable and pleases you. The best model for this is the stony creek look where the mulch is actually large stones with the plants appearing between. As the plants grow, the next row of rocks can be removed to allow for the increase in size.

Q. *Can't we just have the bird relocated?*

A. With the proper permit, this can be possible, but...

If your back yard is within the natural distribution of the Australian Brush Turkey and you are determined to plant a rainforest in the suburbs, your garden is a turkey magnet. All trees, moisture and mulch! There will simply be a long succession of Brush Turkeys. Each one you remove will be replaced. My garden is now much like the pre-European gully it once was. We have our resident turkey* (Son of Giblets) and we like the overall effect. After all, what's a subtropical rainforest without a subtropical rainforest Brush Turkey?

Vegetable gardens and other intensively managed areas need to be caged and if you want a turkey-free garden, it has to mimic non-turkey habitat. Non-turkey habitat is all mown grass, the favoured nesting site of angry Spur-winged Plovers (Masked Lapwings), and a few geraniums. Not my cup of tea.

War of the roses

Some of my wild friends will help themselves to fruit, flowers and foliage in the garden without asking. Sorry, but not being a keen gardener of the European style, I'd never given it a lot of thought until I was invited to Government House in Brisbane to advise what might be done to prevent the resident Ringtail Possums from doing what they were doing to the Governor's roses.

They were eating them. Evidence of their culpability could be seen in the little tufts of reddish fur dangling from some of the thorns. I really don't think it was an issue for the Governor. Clearly it was an issue for the Head Gardener, a passionate man. Wildlife management wisdom of the mid-1980s prescribed a dose of hot English mustard. I should have given it to the gardeners. But I didn't, it was early in my career and I painted it on the rosebuds just as I'd been told to do. In this experiment two important facts were revealed. The first was that Ringtail Possums like hot English mustard and the second was that hot English mustard makes a heck of a mess of rosebuds. The more I experimented the more I came to believe that physical barriers were more reliable than voodoo. I tried hot chilli. I boiled those little red ones in some water in a saucepan and got a result that was a higher octane rating than Tabasco sauce, by a factor of two. I treated a variety of garden plants, fruit and vegetables with it and was never happy with the result.

One slightly different version of the War of the Roses was actually the War of the Pumpkins. A family on acreage had a monstrous pumpkin patch. The children supplemented their pocket money with weekend pumpkin sales on the footpath. It was for them a significant enterprise and I have to say the pumpkins were excellent. Without any formal declaration, one night several of the pumpkins were attacked.

* Sadly, no more. See page 122.

They were about half grown and green. The next night, more of the same. After three weeks or so, about a third of the developing pumpkins had excavations and the tooth marks fitted my mate the brushtail possum. This was confirmed by nocturnal observations. On any one night up to four brushtails were seen munching pumpkins.

I was still in my "deterrent" days, so I tested the full armoury. I painted the affected pumpkins with hot mustard, the special high octane chilli, quassia bark brew and even tincture of asafoetida, a particularly foul smelling liquid that was used in days gone by to treat asthma and other lung related illnesses. I decided that asafoetida would have cured just about anything, or if it didn't, the patient would say anything to avoid a second dose, including "I'm fine now, thank you".

It was difficult to determine whether my paint jobs were having any useful effect, but I concluded that they weren't and the pumpkin patch was beginning to offend. Meanwhile, we pegged bird wire over some of the developing pumpkins. At about the third week of all these treatments, the local brushtails suddenly went off pumpkin. Night after night no new munchings were noted.

I was firmly of the view that the wire mesh worked and everything else probably didn't, but there was no doubt that whatever had caused the sudden pumpkin mania in the possums had gone as mysteriously as it had arrived. Chemical changes in the local food supply? Lots of plants do this to defend themselves from being eaten. It wasn't that the possums had left: they were still seen and heard regularly. No, the possums had actually stopped eating green pumpkins. The other piece of the puzzle was that this family had had the pumpkin patch for about five years and it had not attracted any possum attention in all that time.

In the washup of the War of the Pumpkins, I made the very strong point that I was sure my treatments had not been the cause of the end of the possum raids; that it was a single event that hadn't happened before and might not happen again, although I couldn't explain why. We agreed, however, that the bird wire was a definite winner. None of the pumpkins so covered had a single tooth mark in them.

About five years later, I bumped into the pumpkin lady at the local shops. I asked if they were still growing pumpkins (yes) and whether there had been any repeat attacks by possums (no). What she said next just about flattened me. She'd carefully stored all the bottles of hot mustard, the special high octane chilli, quassia bark brew and tincture of asafoetida and if they ever came back she was ready for them! I was going to mention the bird wire mesh but decided I needed to do a course in Effective Communication instead and looked around giddily for somewhere to sit down.

Lots of things we like to eat will also appeal to wildlife, so we shouldn't be too surprised if that happens. We need to be rational about it, though. If the parsley is being cropped to the point that you have no parsley for yourself and growing it yourself is important to you, grow your parsley in a cage. An old cocky cage is perfect. There is also the issue of acceptable loss. Professional fruit growers lose an average of about 30% of the total crop each year to a variety of natural things. Birds and flying-foxes are the principal vertebrates involved but it is the much smaller forms of life that are the biggest challenge. Viruses and invertebrates (especially moths and beetles) are pretty formidable adversaries. Growers don't like the loss, but it is a fact of life and minimisation is an economic issue.

Wildlife-proofed herb garden in a cocky cage.

In the back yard, you can do the same. If it's really, really important, don't waste valuable emotion raging at the beasts, prevent! I've seen some wonderful sights on my rounds.

People of Mediterranean origin, particularly, love their fresh figs. The best figs are produced on a fig tree grown inside a big wire cage.

A bird-proofed fig tree.

Bird and flying-fox proofing for fruit within reach.

It works. Possums and flying-foxes will eat pawpaws. The trick with pawpaws is to pick the pawpaw the day **before** the possum wants to eat it. If you're adamant that the only pawpaw worth eating is one that has fully ripened on the tree, then the icecream bucket brassiere is for you. This one comes courtesy of my father Owen who is indeed a fancier of tree-ripened pawpaws. Punch a couple of holes near the rim of a two litre plastic icecream container and slot some elastic through the holes, tying it with enough slack to go round the tree with the container over the pawpaw that is next to ripen.

Leave the pawpaw to the sun during the day (possums and bats are nocturnal) and simply, at dusk, slip the pawpaw bra over the fruit, making it safe from possums. If the tree has grown too high to do this easily, it's time to cut it off at about two metres high, put a suitably sized tin over the cut trunk, and shoots will rapidly develop and become thick trunks bearing fruit at a convenient height.

Bird wire tacked along the top of the passionfruit trellis and draped down over the vine will protect the passionfruit from possums. A vegetable garden completely surrounded by a six feet high wire fence and topped with a wire roof will ensure that everything inside is safe from vertebrate raiders. Insects and viruses are still another matter.

I'm less enthusiastic about the loose nylon netting tossed over the tree. It seems to be great in preventing bird predation, especially if it's the white version which they can see clearly but at night it's a different matter. I've spent a good many hours helping to untangle unfortunate flying-foxes from the nets and more than half of these are fatally injured in their frantic efforts to escape. I've also seen bites on people who have attempted to free a bat from floppy netting, and bites from bats in this day and age are a very serious matter requiring rabies treatment if the bat tests positive.

It's a lot more work, but the rigid wire mesh cage is a long-term, highly effective protector of figs and other precious fruits. There is another method that combines wire netting, steel pickets and flexible polypipe. The pickets are driven into the ground on opposite sides of the tree. The polypipe is slid onto the picket, bent over to the picket's opposite number and slid onto that, forming a high arch over the tree. This is repeated on diagonally opposite pickets with the two arches of poly-pipe forming the frame on which to attach the wire netting. Works quite well, but personally if I really treasured my backyard fruit tree I'd have the permanent cage and to heck with aesthetics.

A pawpaw bra in action.

Eggs-cavations

Q. *A big grey lizard has just dug a hole in our gravel drive and laid eggs. They'll be run over. Can we shift them?*

A. There's no need to.

This is not an uncommon enquiry. In spring, most species of reptiles start courtship, mate and, for the egg-layers lay their eggs. Not all reptiles lay eggs; some, like the Blue-tongued Lizards, give birth to live young, but the most commonly seen egg-layers in the back yard are the Bearded and Eastern Water Dragons. The females excavate an egg chamber in suitably moist soil, install the eggs and refill the hole. Two to three months later, the baby dragons hatch, push their way to the surface and head off to battle for their precarious chances of survival.

Strangely enough, even though the eggs are only centimetres under the surface, they are perfectly safe from everything except predators and flooding. Predators are few, but if rain floods the area and the ground is saturated for any length of time, the eggs die. This doesn't happen often and it seems that mother dragons are innately adept at choosing a sensible site.

The concerned backyarder who likes lizards fears the eggs will be harmed if left buried on the driveway. The physics of a narrow hole, a fat tyre and a brief passage above, all combine to make the eggs safe. The soil above the eggs would need to be operated on by one of those huge, noisy, vibrating compacters before the eggs were damaged. Digging up the eggs and incubating them is certainly possible but it has grave risks. Rolling the eggs around is known to cause the death of embryos, so it's best all round to simply leave them where they are. They'll be fine if they're not dug up.

Alan and Stacey Franks

Bearded Dragon.

Q. *Will we see the lizards come out?*

A. It's possible but very unlikely.

One of my listeners has recently reported seeing the hatchlings, but he was very fortunate. Firstly the smart babies emerge at night when predators are fewer. They are still at risk, though. I once found a road-killed Eastern Small-eyed Snake and, being incurably curious about anything in Nature, I excavated the belly and found a baby Bearded Dragon inside. Small-eyed snakes are wholly nocturnal so there are still risks at night but nothing compared with daylight hours. Secondly, the incubation time is so variable (between two and three months depending on the average temperature over that time) that being at the right place at the right time is a bit like rolling dice. But if the opportunity arises, give it a go. Look daily in the very early morning from about seven weeks after laying.

Q. *We saw a big goanna up a tree in our back yard and it looked as though it had dug a hole in the termite nest on the tree. What was it doing?*

A. Laying eggs.

One of the most fascinating aspects of the reproductive biology of some monitors (goannas) is the female's use of termite mounds to assist in egg protection and incubation. This has only relatively recently been reported. It seems that mum Lace Monitor, *Varanus varius,* climbs up the termite mound, tears a section off the side, digs out a cavity large enough to accommodate her clutch of eggs and deposits them therein. The termites frantically repair the damage and, in doing so, cement the eggs in and cover the breach so that in a few days only the scar of the new mud indicating the repair remains.

After a period of about three months, the babies are full term and need to get out. Mum returns and digs the babies out!

I cannot claim to have witnessed this miracle of Nature (for miracle it is), but I have on more than one occasion seen Lace Monitor eggs encased in parts of an arboreal termitarium that has fallen from a tree during a storm, so there is no doubt about the habit. On one occasion, the eggs were only days off hatching and babies that appeared were spectacular. They were a deep blue-black with pink tinges, white bands and little circles. Like all snake eggs I've seen, monitor's eggs are glued together with an adhesive that rivals the modern super glues. Any attempt to separate them is likely to tear the parchment-like shell instead. Upon hatching, another miracle of Nature occurs. One of the components of the amniotic fluid which surrounds the developing baby is the solvent of the glue that binds the eggs.

As the hatching young slice through the shell, fluid spills out and dissolves the glue and the eggs miraculously fall apart allowing everyone to escape!

Steve Parish

The resourceful Lace Monitor.

Nowadays, whenever I see an adult Lace Monitor on the side of the road, a victim of motor traffic, I worry. I've often stopped to see if I can determine the gender. Is it mid to late summer? Is it a female carrying young? It's a pretty forlorn hope that I will be able to rescue any babies, but it's always worth the effort.

Q. ***We were weeding along the fence line and came across some rubbery little white oval eggs that bounced. What are they?***

A. They are likely to be skink eggs.

There are more species of skinks in Australia than any other reptile family and they are a complicated bunch. Skinks range from the giants of the big black rainforest mob called Land Mullets and the familiar blue-tongues to weeney little things without legs and everything in between.

It is very difficult to define a skink in a way that separates them from all the other lizards but funnily enough most people have a pretty good idea of what a skink is. For those who are uncertain, the best place to start is with the smallish brown jobs that dart around the garden or climb around the fences.

Skinks tend to lay in damp spots in the spring and the babies can be seen from about early January onwards. If you look closely at the adults in November and December you can often see the egg bulges in the females before they lay.

Q. *We've seen a tortoise digging in our back yard. We didn't get too close but we think it laid eggs. Would that be right?*

A. Yes, indeed. There is only one reason that a fresh water turtle ("tortoise" is usually reserved these days for the totally terrestrial chelonians of which Australia has none) would go to the trouble of digging.

Like their reptilian cousins the dragon lizards, all female turtles are egg-layers and usually come out of the water to deposit them in a safe place. If the eggs are laid in water and remain in water for any great length of time they are unable to develop and subsequently die. The hind feet or flippers are the nest diggers and generally once the egg chamber has been dug out and laying commences, nothing will stop the rest of the process. One of the most famous Australian examples of turtle nesting occurs annually and very publicly at Mon Repos beach near Bundaberg in Queensland. This is something to be experienced at least once in a lifetime. They are the giant marine turtles rather than the much smaller freshwater ones but the principles of excavation and oviposition are the same.

Steve Parish

A lovely Long-necked Turtle.

Q. *But the creek and the dam are a long way from the house yard. Is this normal?*

A. Yes. The mechanism that tells the female turtle to choose a site to dig is a bit of a mystery, but she will keep trekking away from the permanent water till she "knows" the site is about right. And she doesn't always get it right.

Lots of eggs never hatch and, of course, the perils for hatchlings are huge, but the strategy of these ancient reptiles is to live a long life, lay a lot of eggs and let Nature do the mathematics. One Broad-shelled River Turtle, *Chelodina expansa,* was observed digging at least three kilometres from the nearest aquatic environment. Quite a hike!

Q. *How long before we see the babies?*

A. Anything from three or four months to two years, depending on species and location, and only if you are very lucky.

This is the hardest question to answer. There are records of freshwater turtle eggs hatching after two years in the cooler regions of its range, but in the tropics and sub-tropics under average conditions, eggs that are laid in early spring hatch in mid to late summer.

The Broad-shelled River Turtle, already mentioned, lays before winter and the eggs hatch in mid summer so their incubation is much longer than most other species. The hatchlings (if indeed they hatch) will usually emerge at night and it seems the nest site may need to be rained on to enable them to get out of the hardened soil, so nothing is too straightforward.

Q. *A couple of weeks ago we had a load of topsoil delivered. We're ready to use it but these tiny little birds have dug a tunnel in it. What are they doing?*

A. Nesting.

Pardalotes and the way they nest is one of the natural world's great wonders. In late winter, early spring the males start drilling. Their normal nest sites are dry gully banks and sandy slopes, but even quite flat terrain can be used provided it can be excavated and will not cave in.

Although Pardalotes are remarkably industrious for such tiny birds, if someone or something is kind enough to drop a tonne of easily drilled soil in their breeding territory, they will gratefully run a nesting burrow into it rather than the tougher substrate nearby.

What is most bizarre is that their ancestral nesting genetics still appear to be fully intact.

Pardalote nest holes.

Although they dig a nice secure tunnel sometimes up to a metre long, they then go to all the trouble of carting in a great load of grass stems and, at the end of the tunnel, scrape out a nest chamber and build the typical neatly woven cup-shaped nest you usually see in the fork of a tree!

Q. *When can we use the topsoil?*

A. That is the big question.

My feeling is that it's safest to wait until all the nesting activity is over, but that could be months. Pardalotes lay a large clutch. The eggs hatch in only 14 days and the babies develop rapidly. From beginning to end the whole process can be as little as a month. However, if conditions are good, they can do it all over again.

When nesting is in progress, the adult birds visit the nest constantly, all day long, so an hour or so each week spent watching what's happening would profit you and the Pardalotes.

Take your trusty binoculars and watch the tunnel entrance from a concealed point 10 or 15 metres back. If all is quiet, nesting is probably over. However, if you misjudge and find that nesting is in progress, contact the local bird carers. I've seen quite a few baby Pardalotes successfully hand-raised and released to the wild after an accidental invasion of privacy.

Poultry excuses

Chooks, ducks, geese and turkeys are all vulnerable to a variety of wildlife. The first Spotted-tailed Quoll (now endangered) I saw at close quarters was a large and magnificent male that had been shot in a chook house. I think the first Emden Goose I saw (I didn't know what breed it was at the time) was slowly disappearing down the front end of one of the biggest Carpet Pythons I'd ever seen. I arrived at one chook house in response to an SOS to see the first Square-tailed Kite (a rare species) I'd ever seen at close range and it was full of Rhode Island Red and looking a bit sheepish. I've consoled chook owners who have waged a constant battle against the local Lace Monitor for eggs. First in first served.

In all these cases, the thing that was seriously at fault was not my friends; they were victims of an elaborate set-up. It was simply the inadequate design and construction of the chook house. Young goshawks, having been given their marching orders from their parents breeding territory, are forced to fend for themselves.

If, during their exodus they stumble onto a mob of easy-to-catch finger-lickin' good chickens, they will avail themselves thereof. Probably the worst chook house design is the small pen from which the chickens can't get out but to which the goshawk can simply drop in – for lunch. A good rooster helps. There's nothing quite like a good rooster – just ask any hen – but a good roof is even better. A roofless chook run is asking for trouble. Free ranging chickens have a better chance but they are still vulnerable to predation.

Alan and Stacey Franks

A Carpet Python with "missing" chook.

Maximum security

At night, the roost of a good poultry house comfortably accommodates all the residents without any having to roost under the others, and is completely enclosed in a wire mesh of not greater than 12 mm square. That's half an inch. At a greater mesh size, Carpet Pythons of a size capable of swallowing bantams can get in. The maximum security area should also contain the laying boxes. If the hens are kept in until late morning, the eggs will be in them and they will be safe from the Lace Monitor, which, once it has discovered an egg supply, will visit daily during the warmer months for its share. Often its share means the lot!

Keeping poultry also means rats and mice and rats and mice mean snakes that eat rats and mice. Favourite foods of the Eastern and Western Brown Snakes are the introduced rats and mice. Maximum security means pulling on the gumboots to go into the chook run and being able to see clearly where you put your hands.

If the poultry is to be allowed to free-range (and, if you're going to use birds in this way, free ranging fits better with my philosophy) you need to supply them with ample shelter sites around the yard to in case of a goshawk raid. Sheets of iron or planks of timber on brick blocks high enough to dive under when alarmed are useful. And a good rooster. If one of your hens is grabbed by a goshawk, the rooster will not allow it to calmly pluck and eat the unfortunate bird: it will charge the intruder and see it off the premises. If it doesn't – get a better rooster.

Ian Morris

Western Brown Snake.

Raiders of the lost park

The first practical lesson in natural history I ever remember from my father was when he showed me little conical pits in the back yard and explained that these were dug by bandicoots. I was very young but the memory is quite vivid, which probably means I have it all wrong. Nevertheless, it's my version of events. Much later, I was to discover that a lot of people were far from casual about the little conical pits dug in their lawns by the bandicoots! Some became apoplectic at the sight of those little conical pits. In 1984, I found myself on the receiving end of bandicoot rage. It is an interesting phenomenon and one which was poorly understood in the early 1980s. The standard advice was to tell the caller to lay a 150 mm strip of blood and bone around the perimeter of the yard to deter the raiders or, if the customer demanded further satisfaction, to provide him (rarely was the complainant a woman) with a permit to allow the protected bandicoot to be trapped and relocated.

I confess that I never applied the blood and bone remedy in the prescribed manner due to cost, but a limited trial of a few metres across a track used by the bandicoot caused it to pause... sniff... eat some (I think – it was very dark) and move forward to the lawn whereupon it searched for a good place to dig.

Bandicoot diggings.

Enquiries of mammalogists bore little fruit. I was unable to find out what the diggings were all about. It was obvious that the bandicoots were digging for food, but just what food? Some thought it was the juicy grass roots. I decided to find out. There was a file on bandicoot permits and I was immediately struck by the limited distribution of the dates. With one or two exceptions, all the permits were issued from April to August. So this was an autumn to winter thing.

I tracked down almost fifty people to whom bandicoot permits had been issued. The conversations were very interesting. Only one of the permit holders reported catching a bandicoot, which I counted as a good thing, and only one of the permit holders had experienced the phenomenon twice and there was almost 15 years separating the two occasions. Clearly this was not a huge issue for the suburban community as a whole.

Securing permission from my very next bandicoot complainant, I sat on a lawn in the chilly dark for several nights to see if I could discover the bits I wanted to know. The food is scarab beetle larvae. Green keepers call them "white grub" or "curl grub". As children, we called them witchety grubs but they are not. Witjutu grubs are cossid moth caterpillars that tunnel among the roots of the Witjutu bush in central Australia.

Steve Parish

Northern Brown Bandicoot.

The scarabs responsible for the ones I was looking at were all beetles we call Christmas Beetles, those beautifully metallic green and gold beetles that sometimes irrupt in large numbers and crash into walls and windows at night. The grubs, or larvae, actually feed on the root systems of the grass. The concern being expressed about bandicoots damaging the lawn was totally misplaced: although the diggings were unsightly, the bandicoots were, in fact, preying on the lawn predator. My observations on that lawn and a few others suggested that very few of the larvae escaped the notice of the bandicoots.

They track them by sound. They can hear the grubs' little jaws chewing away at the lawn roots and immediately dig them up.

You don't have to be Einstein to work out that if you remove a predator that is consuming a very high proportion of prey, the populations of their prey will increase dramatically. Each surviving female beetle can lay more than 300 eggs. This relationship is not to be messed with if you want to have lawns in the future. If the grubs survive, the resulting massive increase in adult leaf-eating beetles leads to dieback in the big trees.

Keen lawn keepers resent the mess the bandicoots leave and fear they'll come back and do it all again. The mess is easily dealt with: give it a quick rake and plug the diggings. Since the bandicoot's target has been consumed, it won't waste time digging another hole there, so no repetition. The fewer the grubs, the fewer the beetles; the fewer the beetles the fewer the grubs. I think the bandicoots are on our side. In fact, I know they are.

There is a really weird thing about humans. Why would you cut down trees that are closest to you and that you can most clearly see so that you can barely see trees a long way away? In coastal Queensland, some people were investigated for killing protected trees that blocked their view of the distant horizon. As a community, we seem to have a love/hate relationship with Nature: some elements are adored and others abhored. I find the happiest people are those who have no prejudice or bigotry. But there are also those who probably just shouldn't venture Further Afield…

All the Rage

Crows

Crows, ibises, magpies, possums, kangaroos, turkeys, currawongs, martins, seagulls, miners, snakes…the list is long. These are the species that are simply not rare enough for their own good.

Steve Parish

Not rare enough to be treasured.

I was in a far western town a few years ago and there was a bit of a soirée passing itself off as a rough barbie and, relaxing with a can of a well known Queensland brew, I was looking skywards at the breathtaking display of Black Kites, or Fork-tailed Kites as we once called them. They're hardly black. In fact they are hardly any blacker than our Pacific Black Duck and it can get into any venue as a brown duck nine days out of ten. But I digress....

The sight was, as I said, breathtaking. You don't see a sky full of gracefully wheeling Fork-tailed Kites in my country of south-east Queensland. "Have a look at that," I sighed to no one in particular. I was immediately startled by the bloke standing next to me who said, "Yeah, 'spose we'll have to start doing something about them soon - hey?"

Taken aback, I said, "Why, what do you mean?"

"Well, that many, they gotta be doin' something, hey," was the reply.

That was when another golden penny dropped. That's it! I thought. That's the key to all this wildlife rage. We've ended up after only 200-odd years of European Australia with a culture that is plagued with a fear of plagues. Real or imagined. Rabbits, mice, rats, locusts, ants. Never people, of course. We still don't have enough of them for our own good. Not much we don't!

Q. *But there **are** too many crows – aren't there?*

A. If, by "too many" we mean too many for us, then that is a subjective and sadly oft-heard comment. But if we mean too many for their own good or Nature as a whole, then the answer is implacably "no".

I would define "too many" as a state of being perilously unsustainable. This can and does happen. In the normally arid country of the north and centre, there lives a native rodent by the name of *Rattus villosissimus*, the Long-haired Rat. The species lives a life of boom and bust. It is normally quite rare, but following unusually high rainfall, this rodent's numbers explode. Feeding at night on the abundant green grass and seeds resulting from the rains, it increases to the point where there is a corresponding increase in most of its predators as well. Given the nature of the dry lands, this can't last. It is not the predators that cause the bust – there are just too many rats for that. It is the diminishing food supply, and when most of the food has gone the rats crash. So do some of their predators, particularly the huge number of young that have been born and fed on the abundant rat. This is a case of "too many", but what could be done about it?

Crows also increase in response to an increase in food supply but much more slowly, and they are constrained to some extent by the protocol of the breeding territory. Even with a big food supply, Torresian Crows do not tolerate another nest in the tree next door.

People often mistakenly attribute a "sudden" increase in the local crow flock to unhealthy mass production, but it is, in fact, an aggregation of young, unattached, non-breeding birds gathering close to the local food supply.

It is true that populations of crows and ravens have increased dramatically in urban areas the world over. The growth in the birds' populations is linked to increased food supply, which has a lot to do with greater affluence resulting in more motor cars, more road kills and greater human food wastage. In Japan's capital, Tokyo, millions of crows are being sustained by the contents of millions of plastic bags containing millions of morsels their crows have come to recognise as the new food.

Q. *Crows aren't native birds – are they?*

A. **Yes, they certainly are.**

Crows and ravens belong to the taxonomic family *Corvidae*. Corvid birds are found world-wide and Australia has five native species. All regions of Australia have at least one species and, because of their habit of scavenging, this is a good thing. They largely feed on garbage which is what makes them so precious and special.

The waste management team.

Q. *But crows chase all the other birds away – don't they?*

A. No.

If crows did that, our bird books would contain only one page by now. My experience is that crows' physical size and loud calls draw people's attention to them and away from everything else.

The anti-crow lobby sees no good in the birds at all. There exist literally thousands of bird lists made by bird observers in all parts of the country. Few of these lists do not include at least one of the local corvid species, but no list I've seen has only crows on it – proof positive that they don't chase other birds away.

Dozens of research projects in Australia and the rest of the world have been conducted in response to the popular notion that crows are responsible for the loss of other smaller, prettier or rarer birds. However, none of these studies has nailed the crow as the culprit. It's always turned out to be the bulldozer, I'm afraid.

Q. *But crows are cruel – don't they peck the eyes out of lambs?*

A. Sick and failing lambs – yes; healthy lambs – no.

To quote CSIRO researcher, Dr Ian Rowley, known the world over as THE corvid guru: "A crow is no match for a healthy lamb and it is those lambs that are already weak from some other cause that are attacked." Dr Rowley's research, which began in the 1960s, resulted in an enormous number of scientific papers on crows in the rural environment and those who doubt my impartiality might like to look in the reference list at the back of this book for details of the two most important of Ian's works on the matter.

I once met a bloke who had studied the Red Sheep in the wild in Iran. It is an endangered species in the wild, even though it is the wild form of every single breed of domestic sheep worldwide. He claimed that the wild ewes avoided predators and protected their newborn lambs during the birthing process by hiding in a dense thicket. This habit provides security, obscuring them from the view of predators of any kind.

I've looked in a lot of sheep paddocks in my time, but I've never seen any thickets in which the ewes could protect themselves and their lambs.

Q. *I teach at a school where there are a large number of crows. Their noise is so loud and constant that it really does interfere with teaching and learning. Is there really anything we can do?*

A. Yes. Since the presence and numbers of crows relate to food and/or suitable roosting sites, you will need to conduct a campaign of food reduction.

In most schools I've been to, waste management is generally not very well conducted. There are inexpensive designs for rubbish bins that will exclude all birds, including crows. Students need to be heavily involved in waste management, or the program will fail. Tossing food scraps around is a real no-no.

Often the school is not the sole source of crow food. Sometimes there is another school nearby, sporting fields, shopping centres, and so on – everyone has to be involved. The best time to start the program is first day back after a long break. This is the time of lowest food supply if the school is itself the primary provider. As the crow food reappears with the students' return, it will take only hours for the crows to discover that the feast is on again.

Q. *Our crows are just roosting in one big tree. Should we have it removed?*

A. No. Cutting down trees, especially big native trees, is not a very positive example for the next generation.

However, crows can be frightened away from a roosting tree by appealing to their neophobia, or fear of the new. Metallic-skinned helium balloons tethered on lines and allowed to drift about in the roosting tree has worked in some cases, but you will also need to do the hard work on the food supply or the lure of the food will rapidly overcome their suspicions and you are back whence you started.

It should also be remembered that the large and dynamic roosts of lots of crows can be very temporary affairs. A lot of the big roosts that people notice and complain about, disappear as quickly as they appeared as the scavenging birds move on to messier pastures. The birds in these roosts are all young, unattached and non-breeding, and mortality is very high. Most don't make it to the next stage of life, while the very fortunate few that do, pair up in an available breeding territory. New breeding territories are being created by urban spread, more cars and more waste.

Q. *Couldn't we just catch them and relocate them?*

A. No. Not only does relocation rarely work, it is expensive.

In the case of the crow, relocation is extremely expensive! The expense is in the catching. The most perfect crow trap will never catch you more than a few birds a week, if that. They are just so smart.

The cleverness is actually neophobia – the fear or wariness of something new or unfamiliar. Once the crowd of crows observe another crow in the trap, with all the attendant stress associated with your grabbing it and wrestling it into a carry cage, you can be sure the trap won't catch another crow for quite a while. In the meantime, the caught and shifted bird has already been replaced from nearby stock.

Q. *What's the difference between a crow and a raven?*

A. A little, but not a lot.

In Europe, crows and ravens are distinguished by the presence or absence of protruding feathers on the throat. These are known as hackles and can be best observed when the bird is calling.

One Australian species possesses prominent throat hackles so it got the name Australian Raven, but beyond that it seems a bit wishy washy. The common names simply developed over time and there is no pressing reason to change them.

The Torresian Crow is *Corvus orru;* the Little Crow, *Corvus bennetti;* the Forest Raven, *Corvus tasmanicus;* Little Raven, *Corvus mellori;* and the Australian Raven is *Corvus coronoides*. As with the "difference" between frog and toad, I've had to stop worrying about it.

Q. *The crows' nest in our back yard seems to have an albino baby in it. Is this rare?*

A. Yes, extremely rare.

Having never seen one, I thought I'd see what the internet might turn up. I could find only one record of a truly albino crow and that was in Western Australia and published by the then WA Department of Fisheries and Wildlife in 1979. They must occur, but the incidence of total albinism must be rare or I suspect there would be more records.

What most people see as an albino crow baby in the crows' nest is in fact a baby Channel-billed Cuckoo (see Cloud Cuckoo Land on page 119).

Heck Lindsay

Fooled by the cuckoo.

Father of the year

Darryl Jones's book, *Magpie Alert – learning to live with a wild neighbour,* is the first publication to deal with this phenomenon in popular language, but it is based on the scientific studies he, his colleagues and students have undertaken over the past decade or so. And didn't we need it! Tens of thousands of Australians are swooped on annually. Some are struck and some are struck severely. Anyone who thinks there are simple answers to the effects of the magpie season is in for a surprise. Rage all you like, for every magpie opponent there are a hundred maggie lovers. What has always astounded me is the lack of respect the magpie's critics have for him, for it is almost always a male that does it.

In exceptional circumstances females will join him or even take over his gladiatorial duties if he is removed, but generally the responsibility for brood defence falls to the male. He's the one with the slightly bulkier build and stark black and white plumage. The female tends to be more lightly built and has grey smudging to her white parts, especially on the back of the neck.

Faced with clear evidence of threats to his wife and kids, the male magpie will defend their lives almost at the cost of his own! Despite this, the popular media rarely laud his loyalty, instead slamming the behaviour as vicious and branding the bird a menace. When humans risk their very lives to protect their family, we pin medals on them. In our society one of the highest honours that can be bestowed on a man who has lived an exemplary life as husband and dad is to be awarded Father of the Year!

Big daddy.

In 1984 when I was faced with my first magpie season as a wildlife ranger, I was astounded at how little information there was in the department of brood defence behaviour. The common assumption was that all male magpies "did it". No they don't! "They are just defending their territory". Not quite!

Darryl Jones's studies revealed that the percentage of suburban male magpies that swooped on humans during the nesting season was considerably less than 10%. In the early 1980s, I estimated about one in twenty and across the landscape, including the urban and rural areas, that's about right.

My big question was, "Why?" Not just why the ones that did it did it, but why the ones that didn't didn't. Genetic variation? Could it be that some were more aggressive than others? The corollary of that is that some must be wimps! Nah! You don't get to the top of the magpie tree if you're a wimp. It's a long, tough trek. You fight your way there. Obviously there was some event or experience that happened to particular males to initiate the behaviour. I called it "triggering".

In the magpie seasons of 1986, '87 and '88, I kept detailed records of all complaints received from the public about "vicious" magpies and mapped them. There was a strong correlation between the places where baby magpies had been rescued and those where new attackers were found. The data weren't showing an absolute match, but there was more than just a hint. Then, in 1989, when a colleague and I were setting up a display of live native animals at a primary school fete in Ipswich in south-east Queensland, a young lady arrived clutching a bundle of grey feathers. She was shaken, her face was ashen and she had long squiggly red claw marks on her cheeks and neck.

There was blood. The bundle was a baby magpie. I asked her if the bird's daddy had got a bit upset and she just nodded. We took custody of the maggie and I went to talk to the school's management.

No, there were no problems with any birds at the school. In fact they were certain there were no magpies at this school. Now, that's impossible. A school without magpies is like a motorway without cars. You build a school and the magpies will come! The reason for this is that schools essentially create a magpie breeding territory. All that mown grassy area? This is the stuff of open woodland – the prime habitat of the Australian Magpie. I pressed the questioning, albeit politely, until I was satisfied that what we had here was a breeding territory which up until now had not experienced brood defence behaviour by the male magpie. When I took the boss for a bit of a walk and pointed out the magpies on the oval, he was convinced that there were indeed magpies at the school, but was adamant that those birds had never been any cause for concern.

Steve Parish

On patrol – alert but not alarmed.

That was the Sunday of the school fete. The next day was the Monday of the magpie's fate. At about 8.30 a.m. the school's big boss was on the phone to my big boss insisting that the attacks on the children that morning had been so serious as to require the bird's immediate removal or execution. Ergo, the rescue of the baby on Sunday triggered brood defence behaviour.

There are other circumstances that will trigger magpie attacks on humans; in fact, I have documented three types, but there is little doubt in my mind that the most common cause is chick rescue. As ironic as it might be, the principal cause of male magpies turning on humans is human kindness. Who, after all, can walk past a defenceless baby magpie stranded on the ground with no one to care for it? Me! At least until sundown.

I've gone back at night and rescued chicks when their daddy couldn't see who did it and everything's been fine. But if he sees, watch out!

Magpies don't like you peering intently at their nest either. I supervised a young naturalist in a study for a conservation badge years ago. He chose the magpie and we designed a project which required him to determine the size of a breeding territory and relate the observations to development of the young. Naturally, he had to make regular observations of the young in the nest, which he did each Saturday morning for six weeks. By the third Saturday morning the male was meeting him at the edge of the park and giving him a biff. The bird took no notice of any other humans that ventured across the breeding territory.

A magpie whose baby had been rescued by a lady pushing a stroller only attacked people pushing strollers. Normally, magpie nests are seven metres or more above ground height. At this distance above walking humans, the magpie is usually unconcerned. Quirks of the built environment can change all this. One female (she chooses the nest site and does all the construction) had picked a lovely big Queensland Blue Gum and fashioned a typical nest about fifteen metres above the ground. All would have been well except for the fact that the tree was growing right beside a bridge, and the bridge had pedestrian paths on each side. People crossing the bridge were coming within a metre or so of the nest. Not only did the male object, but because the potential attackers of eggs and, later, young were so close to the nest, the female was joining in on every occasion.

Cyclists were having a bad time of it as well, with not one but two very distressed screaming magpies buzzing around their heads and flicking their ears as they attempted to cross the bridge and dodge both the birds and the traffic. Tricky one.

There is no doubt that the best defence is to avoid the area for the season, but I've found that a good umbrella works. I remember doing a piece for the television news in a park where there was a triggered male. I strode across the park to demonstrate that it was perfectly safe provided you had the protection of an umbrella. I'd borrowed a female colleague's umbrella for the purpose. I thought I would make a big statement for some of the blokes who used to ring and tell me how terrifying the local magpie was, so this umbrella was all pretty flowers and pearl handle, and I minced over the grass for the delighted camera man. At the other side, I was attacked by a lady who had been attacked by the bird the previous day. Seeing my ranger uniform, she thought it opportune to let me know what she thought of the bird and what she thought of me since it was always my fault too. I just pointed to the umbrella. "Ridiculous!" she snapped. She said it looked stupid. It wasn't raining. I said it was – big black and white birds!

If the problem arises in your own back yard and the kids can't get outside for fear of being dive-bombed, it's time for some serious kindness. Feed them! Yep! It's all in Darryl's book. There is no doubt that tossing some meat and saying soothing things can dramatically affect your triggered bird. You can, with a bit of patience, return him to his previously lovely "wargla ardle oodle" self.

Magpies are not the only Fathers of the Year either. Grey and Pied Butcherbirds are vigorous defenders of their family. In fact the smaller butcherbirds really put the wind up you. Unlike the maggie, which usually (but not always) swoops from the rear, the butcherbirds will come at you head on and look you straight in the eye first. Spur-winged Plovers are a doddle by comparison. They are all hiss and wind. It's a bit scary at first but you can be pretty certain that the plover is all talk. Peregrine Falcons aren't, though. You go poking around their nesting ledge and you are likely to get seriously biffed.

Bush Stone-curlews will see you off, as will Noisy Miners, Pied Currawongs, Spangled Drongos... in fact, when I come to think of it... if you want a real taste of brood defence behaviour, try getting between a male Southern Cassowary and his kids. They can kill. They have.

Spur-winged Plovers – all bluff and wind.

Flying-foxes

Some humans have gone to extraordinary lengths to try to persuade a large group of flying-foxes to camp somewhere else. The first time I heard of the use of a helicopter to move them on was the early 1980s.

The story goes that it cost more than a thousand dollars an hour and although the bats did move along the river away from the place they had no business being, as soon as the helicopter went back to its roosting spot, so did the bats.

More than 20 years later I was astonished to hear that the same technique was going to be used on a large camp in the Northern Territory. You can tell people about the ecological imperative of flying-foxes – forest health and seed dispersal, pollination, etc. – until you're blue in the face, but if the mayor says the bats have to go, the bats have to go! The really weird thing is that it's technically very simple. Flying-foxes roost in large numbers; sometimes very large numbers. I call them camps rather than colonies. For me, colonies are cooperative aggregations of animals assisting each other to make a living. Flying-fox camps are resting places.

A huge influx of Little Red Flying-foxes – all gone two months later.

When I was chasing camps all over south-east Queensland, I discovered that people tended to view bat camps as if they were cities; that is, they were in a definite place and of a definite size. There was Sydney, Brisbane, Melbourne, and so on, always in the same place and consisting of a certain number of bats.

It just isn't so. One camp I saw that would have contained well over a million animals one January was, four months later, completely bat-less. The size of even long-term camps, ones that have been occupied for decades, varies enormously throughout any one year. Such is the certainty that some people have about bat camps, that one famous Batty Boat Cruise up the Brisbane River in the late afternoon to a very well known "permanent" camp arrived to find a complete absence of bats.

Bat camps can cause hysterical behaviour in anti-bat people. Apart from the hilarious episodes with the helicopters, I've witnessed fires lit under the resting animals, stock whips being cracked up and down the camp and gangs of pre-dawn crazies armed with pots and pans to bang together to deter the bats returning from their night's foraging.

Bat rage begins when the first of a big group arrives at a site that has been previously unoccupied, at least in living memory, or when an influx of the highly nomadic animals swells existing numbers in a camp causing it to spill into back yards, or public parks. The media pounce on opportunities to show the aggrieved human faces. I've yet to see a modern journalist resist the temptation to use the terms "gone batty", "goes batty" or "sends batty" on their first flying-fox camp assignment. For them the terms are highly original and very witty.

Myths and legends about bats abound. Although there is only one single species of bat that feeds on the blood of mammals (the other two restrict themselves to birds), the legend of the vampire looms large over the whole Order. It is true that bats in general are host to viruses that can be harmful, even fatal to humans, but the hosts of viruses that have killed more humans than any other group of animals are actually the birds. I am told that most, if not all of the viruses that become known as influenza, originate from birds.

I was peering through binoculars on the perimeter of a flying-fox camp counting the bats in vertical columns to get some idea of the species' composition as there were both Grey-headed Flying-foxes and Black Flying-foxes roosting together and we were trying to estimate the number of Grey-heads still alive. A young bloke who was a bit interested in what I was doing sidled up and asked me, in a very soft voice, if it were true that bats defecated through their mouth or words roughly to that effect. He may have used another word for mouth. I'd actually forgotten this one – I hadn't heard it for such a long time and thought it might have been extinct. The myth is probably due to bats' natural resting position which is the right way up for them but the wrong way up to us. In fact, bats are incredibly tidy when it comes to bodily functions. They swing their thumb hooks up to their hanging branch, drop their legs down, complete the necessaries in a flash, shake the lower end vigorously and, without so much as a speck left on their nether extremities, swing back down to the right way up! It works for them.

I told my youthful enquirer that, no, bats didn't defecate through their mouths and in fact the only species I knew which tended to do something like that was the one he was looking at, which happened to be me. Fortunately he got the joke and as a result warmed to the subject. He listened to all my points in relation to the positive contributions made to his world by bats and we parted with both of us better off.

Flying-foxes do have a distinctive aroma. A whole camp has a very distinctive aroma, which is amplified during rainy periods. I happen to like it but not everyone does. They are also very vocal. A huge number of flying-foxes suddenly arriving in the gully in the back yard one morning is alarming for some people. For others it is a total joy. Some of the fiercest bat supporters I've met live on the edge of a big camp. They can be very valuable. Plenty of wildlife tours include bat camps on the itinerary. Why not? The camps are always pretty lively. If it's late in the year most of the mums have babies and care for them in a very intense and human fashion.

But back to the "problem" camp. The mayor's announced that the bats have to go. What to do? Well, have you ever noticed that they always hang in trees? If you decide that the bats have to go, you are also deciding that the trees have to go and if the trees can't go then the bats stay. It's a simple as that. Try to remember that, being nomadic, the bats move and most of the noise coming from the anti-batters will subside once the bats head off. There is no doubt that the best and most effective method of flying-fox management has more often than not been **nothing**.

Cane Toads

There is no doubt that the Cane Toad is spreading, but this isn't news. The toxic amphibian has been spreading since 25 June 1935 when it was released into the Queensland wild. Funnily enough, 1935 was the year of the big sugar conferences held in Brisbane and Bundaberg to try to solve the most pressing problems of the sugar industry. And what was the most pressing problem of the sugar industry in 1935? It may surprise you to know it was the same problem the industry had had since the late 1890s – a huge world glut of sugar combined with low global prices. Sound familiar? Why an economic entomologist would be releasing a toxic toad into a sea of sugar beggars belief – but there you go.

And the toad began its march and is still marching. The prospects of a biological control are presently fairly poor. Anything lethal that could seriously damage the toad is very likely to be dynamite in the Australian environment. We are stuck with the toad at least for the present. While it has been fashionable to derogate this amphibian as ugly or revolting, it is a great frog in its states of origin. In fact, it is considered a threatened species in the southern parts of the United States where it is native, and fears are held there for its future.

Australians need to be careful in being too vociferous about the Cane Toad's physical appearance, as there are lovely native Aussie frogs that make the Cane Toad look like the handsome prince. *Myobatrachus gouldii,* Turtle Frog of south-west Western Australia, is a case in point. It is so named because of its resemblance to a baby turtle that's had its shell ripped off, and no matter how you view the Turtle Frog, "handsome" is not the word that immediately springs to mind.

Cane Toads and egg string (circled).

Anyhow, looks don't come into it. A whole swag of foolish introductions have been visited upon our once great land and it is far more important to stop doing it than to tilt quixotically at windmills. With the exception of the European Rabbit, most of Australia's Top 100 urgent terrestrial environmental problems don't have backbones, they have leaves.

It would be wise, however, if the States now home to the toad – Queensland, the Northern Territory and New South Wales – were to legislate and regulate to prevent the toad's spread where this is preventable. Offshore islands are the places I see most at risk and they can be protected with a bit of care and attention.

I was born into a Cane-Toad-free Brisbane. Well, almost. The first Cane Toad specimen lodged at the Queensland Museum from suburban Brisbane was from the Morningside railway station and was registered into the collection on 10 December 1945. The nearest wild toads at that time were well north of there, probably even north of Gympie, so this individual was not part of the wild, expanding population.

This lad was a traveller. It just so happened that an awful lot of military equipment was being brought south by rail just after the Second World War and much of that equipment had been loaded onto the trains in places like Cairns and Townsville where the toad had been established since the mid 1930s. The first half dozen toads from the Brisbane area lodged in the Queensland Museum are all from within a hundred metres or so of the north-south railway line.

Accidental stowaways were assisting the spread as were dams and ponds. In the early 1960s, gardening practices were being revolutionised in Brisbane. The all green lawn with a few gerberas was being replaced by native gardens heavily mulched to conserve water and provide the cooler sub-tropical look. The first branch of the Society for Growing Native Plants was established in The Gap, one of Brisbane's western suburbs, and it was there in the late 1960s I saw my first Cane Toad. By the mid 1970s, the species was well established all over Brisbane, both north and south of the river which is a substantial barrier for mass migration even though toads have occasionally been recorded as making it across.

It struck me that the rapid occupation of Brisbane's suburbs might well have been assisted. One of the favoured garden mulches commercially available in the 1960s was exotic pine bark and all the exotic pine bark suppliers were north of Brisbane and well within the toads' expanding range. I think we helped it along. I think we can do a lot to prevent our helping it along as well, which is why I see those off-shore islands as important reservoirs of species vulnerable to the toads' toxicity and predation.

It is also worthy of note that Queensland and Australia have lost six species of native frogs over the past 25 years and all six disappeared from relatively pristine National Parks and from habitats that are and have always been completely free of toads. A microscopic fungus called *Batrachochytrium dendrobatidis* originally from South Africa got them but how exactly it arrived in Australia is not known. This tiny fungus, deadly to frogs outside Africa, is known to be associated with the African Clawed Frog, a charming little frog which has for many years been used as a pet and was once used extensively to determine human pregnancy.

Toads do not eat eggs, tadpoles or adult Australian frogs. They do eat a lot of other things which is a worry from an invertebrate conservation perspective but they don't eat our frogs. At least, none of the thousands of stomachs that I and friends looked into years ago showed any signs of native frogs. In fact the only amphibian I ever saw inside a toad was another, much smaller toad.

Wherever the toad is now it will probably be for a long time and there won't be any magic bullet to solve the environmental issues. I'm not against investigating possibilities, but recent suggestions of possible viruses send shivers up my spine and make me wonder if the people making these suggestions have any idea what the original problem was.

But what good are they?

I've often been caught insisting that nothing in Nature is useless. This is a bold statement and has attracted many challenges over the years. What, for example, is useful about mosquitoes? At least four species of Anopheles are the vectors that carry malaria, and malaria still kills more humans each year than any other single disease. In Australia, there are a number of viral diseases such as Dengue fever, Australian encephalitis, Ross River virus and Barmah Forest virus that cause terrible suffering and sometimes death in many humans each year. But this is not the sole role of mosquitoes in the natural or the human world.

The general assumption is that all mosquitoes bite (wrong) and that all bites can cause disease (also wrong). All male mozzies, for example, feed solely on nectar – they're lovely little blokes. It's the females that need a blood meal in order to produce viable eggs.

But there is much more to mosquitoes than stings and diseases. In mangrove ecology, mosquito larvae, or wrigglers, are a huge source of food for newly hatched fish. If you could wave a magic wand and rid the world of all mosquitoes, it is likely that you would also put an end to life in the oceans and not just the diseases mentioned.

The web of life can sustain a fair bit of damage before it collapses altogether, but given enough damage, it will collapse. This is the sort of thing a range of very famous people the world over have been tirelessly trying to get across to the rest of us for years.

Q. *So what good do possums do?*

A. They protect trees and a few other things as well.

For anyone who has had the famous possum-in-the-roof experience, combined perhaps with a bit of possum predation of their herb and vegetable garden and a "stampede" of possum boots across the roof in the early hours, this is a very good question.

But when you look past the irritations experienced by some city dwellers, there is a very different, very fascinating story. (It is also a very different story in New Zealand, but there the brushtail possum is an introduced animal that has done enormous damage, as is so often the case with introduced animals.)

In Australia, brushtail possums protect trees by eating the semi-parasitic mistletoe (*Loranthaceae* and *Viscaceae* families). Evidence for this can be seen in the large, usually smooth-barked eucalypts, or gum trees, that are dead or dying from the effects of mistletoe overburden. Most of the trees that I see in this condition are in situations where possums simply are unable to survive.

No brushtails, no tree protection.

Look along the sides of large arterial roads or motorways for big dying trees dripping with clumps of rusty foliage. These clumps are individual plants of native mistletoe. Most are "planted" there by a tiny bird, the very attractive red, black and white Mistletoebird. The bird feeds on the mistletoe seeds, which are sticky and sugary and rich in all the nutriments the bird and its young might need. It passes the seeds in its droppings onto horizontal branches of smooth-barked trees where they will have a good chance of growing.

The bird plants more than it will need, because it "knows" that most of the plants will never make it to the flowering and fruiting stage. They will be eaten by brushtail possums. If the possums aren't there, all the plants will grow. This actually doesn't suit the bird either. If all the plants do grow, the host tree will not be able to support them all and the host tree will die. As the host tree dies so do all the mistletoe plants and the bird is deprived of the very resource it sought to promote.

Then there are the *Delias* butterflies, the jezebels or Union Jacks. Their larvae feed only on the leaves of the mistletoes, and the adults in turn have a separate role feeding on nectar-bearing flowers and providing, as a result, a pollination service to those plants. Some of the jezebels are winter fliers and are the only butterflies that can pollinate winter-flowering plants.

It's complicated! It sounds like a long drawn bow but without the possums, the trees could go, without the trees the mistletoe goes, without the mistletoe there's no jezebel butterflies, no pollination service, no Mistletoebird and it all comes tumbling down.

There's a hole in the bucket, dear Liza. And the bit missing in this particular bucket could be the brushtail possum.

And why isn't the possum there? Usually it is because of high volume, high speed traffic. This is why the most striking examples of the affected trees can be seen alongside major roads. The collisions between the possums and the cars are so frequent as to make survival in the immediate area simply impossible.

Pull over, get out of the car and walk into the nearby bush and gradually the number of mistletoe plants per tree decreases the further you move into safer territory away from the heavy traffic. There will still be some mistletoe plants but the incidences of trees dying from overburden will be few to none. Here the balance is restored.

It surprises most people when they are told that the so-called Common Brushtail Possum is now extinct in about a third of its pre-European range. That's a 30% loss in 200 years. Pretty sobering, that.

It's been my experience that if you dig enough into the life history of any organism you find a connection that links two or more parts of the web of life together. After all, that's how the web was built in the first place.

A Wild World

Myths and fallacies

Lemmings commit mass suicide.

While this has nothing to do with Australian native wildlife, it is a great example of how a myth relating to a wild species can be created and perpetuated by misinterpretation.

Lemmings are an Arctic species of rodent and have the ability, in especially good times, to dramatically increase local populations. Overcrowding can then cause large migrations and during these events some individuals mistake very large bodies of water for small streams. Instead of setting out to cross a trickle, the unfortunate animals find themselves in a marathon swim beyond their endurance and drown.

The scenes in the 1958 Disney film, *White Wilderness,* were shot in Canada and were all staged by camera crews using a few dozen Lemmings that had been purchased from native North Americans. The seemingly careless disregard for truth in the portrayal of the animals' behaviour has given rise to a widespread belief in this myth. In reality, rather than committing mass suicide, overcrowded Lemmings are more likely to wage lethal war on each other for space, which, when you think about it, makes more sense, unfortunately.

Birds mate for life.

There is a common belief that many bird species mate for life, but the reality is that these species of birds maintain fidelity to their partner till death them do part. Upon the mortality of one, the other rapidly finds a new mate. In the case of the Australian Magpie, the period between losing a partner and gaining a new one can be only a matter of hours. However, they don't choose a mate: they take full advantage of a territory vacancy and are perfectly happy to accept any occupant of the opposite sex. The goal is to gain a breeding territory. Much less complicated, isn't it? Some people prefer to believe, particularly of swans, that upon the death of a partner, the other remains widowed for the rest of its life. It is a wonderfully romantic notion, but a myth.

Steve Parish

Like Mute Swans, Black Swans "mate for life".

Harmless snakes keep venomous ones away.

I have been asked many times whether it is true that the presence of a Carpet Python or Green Tree Snake will ensure the absence of venomous ones. The answer is of course "no", but I have always wondered whether this myth originates from a desire to protect the harmless species. There is no doubt that the Carpet Python is the most efficient and effective rodent control that money can't buy. Killing or removing Carpet Pythons even from residential sites is extremely unwise. The potential threat to pets (birds in cages, kittens, chooks, etc.) can be easily managed by providing the pets with security at night in the form of an enclosure with the appropriately sized mesh. But, I suppose, if some people believe that a Carpet Python will keep the Taipan away they might be more likely to leave the python alone.

A second basis for this myth may derive from the reality that some pythons actually prey on venomous species. Black-headed Pythons and Womas, both large snakes mainly of the arid interior, will capture and devour any reptile they can overpower, including the potentially huge and dangerous Mulga Snake. Elapids (front-fanged snakes) and pythons (solid-toothed constrictors) are unrelated, have no connection or association in Nature except the predator-prey one just described, and logically predators and prey are always found together.

A rather rare variation of this one was told to me recently by a lady who clearly believed that a dead Carpet Python's ashes strewn about the garden would keep the poisonous ones away.

Snakes always travel in pairs.

In other words, where you see one snake, there will be another. The sight of one snake means simply that you have seen one snake. There may be others around, but there is no relationship between the snakes themselves: they are solitary animals. Where a particularly good resource, such as food or water or a receptive female, is available, the presence of more than one snake is very likely.

A winter shelter site can contain more than one snake simply because it is a great place for the snakes to spend the torpor – that state of reduced activity some undergo in cold weather. There was a case in south-east Queensland where a shed roof contained more than thirty tree snakes but it was winter and shelter sites were few and far between. The real point, however, is that the snakes were attracted to the shelter site not to each other.

Steve Parish

Whipsnake.

Harmless snakes and venomous ones interbreed.

This total myth is still very popular despite the fact that not one example of this "phenomenon" has ever been produced. And never will be. It is a belief born of the notion that different snakes are different breeds rather than different species. The more unlikely the myth, the more popular it is – for example, that pythons interbreed with taipans producing a terrifying result more deadly than either parent. These two snakes are not only completely different species, they are actually from two different taxonomic families.

Cats and dogs are both of the taxonomic order *Carnivora*. The cat belongs to the family *Felidae* and the dog to *Canidae*. They do not interbreed. All snakes are of the order *Reptilia*, but there remain great and incompatible differences between pythons and taipans. Pythons are from the family *Boidae* and taipans are from the family *Elapidae*. Obviously if someone can show me a cross between a dog and a cat, I'll have to rethink all this.

One of the biggest problems with snakes is simply that everyone is an expert, especially after a few drinks. Most snake experts can be found in the public bar of any hotel a few hours after opening time. It is unwise to challenge their authority on the matter. They can be found at parties, as well. I was once woken at about 2 a.m. by a phone call to settle an argument about the well-known deadly snake crosses and the caller took a great deal of convincing. Eventually, I offered a cash reward of $10,000 for one of these snakes, although I had nothing like $10,000 to spare at the time.

That was many years ago and I still await a claimant. There are quite bizarre cases of hybrids being produced in captivity, which tends to muddy the waters, but they are all from parents that are at the very least members of the same taxonomic family and most often from the same genus. The so-called "Liger" and "Tigon", hybrids of Lion, *Panthera leo,* and Tiger, *Panthera tigris,* are probably the most extreme examples of unnatural hybridisation and people who arrange these things deliberately in zoos should be severely dealt with in some appropriate fashion, for example a week in a small enclosure with the adult male Lowland Gorilla.

Rubber or plastic snakes will keep birds away.

I once read a published letter that asserted that a plastic snake would keep birds away from a particular area for years, even after the snake had been removed. Following this logic, there shouldn't be any birds anywhere in Australia given that all parts of the continent are occupied by at least one species of serpent. I tested this myth some years ago. I got hold of two very realistic-looking snakes and set them up in my bird feeder surrounded by plenty of seed. The King Parrots and the Rainbow and Scaly-breasted Lorikeets just walked all over them. The local Noisy Miners didn't seem to recognise them as snakes, I suspect because they didn't move, and I further suspect that Noisy Miners wouldn't recognise and react to a real snake either if it lay perfectly motionless.

The sight of a snake will certainly cause flight in many bird species and in others can immediately lead to mobbing, when everybody gathers to give the snake a hard time. Many of the gregarious honeyeaters will do this. The Noisy Miner has special calls to warn other members of the extended family of the presence of a predator. The call announcing a bird of prey is different from the snake warning and good birders can tell the difference.

Snakes steal eggs.

There are no specialist egg-eating snakes in Australia although there are apparently a couple of authentic cases of Carpet Pythons having swallowed hen eggs and even one of a Carpet Python that had swallowed one of those china eggs people used to put in the nest boxes to encourage the hens to lay. That always struck me as funny because I had chooks at one stage and all the girls laid eggs flat out, no china egg necessary. Perhaps the china egg manufacturers were having a bit of fun at our expense. Carpet Pythons do not go after eggs. However, it is conceivable that, after being attracted to the chook house by the scent of roaming rats (the usual reason), then latching on to the scent of a live warm bird (the chook), and having missed the chook as she shot out of the nest box, and being left with just the scent of the warm chook… it is quite conceivable, as I say, that the swallowing reflex may take over and the snake eats the eggs in the box still warm and smelling all chook-like. What I do know is that all my attempts to feed eggs to Carpet Pythons failed. Eggs that persistently go missing from the chook house are more likely being taken by rats, goannas (monitors) or crows.

Steve Parish

The Red-bellied Black Snake never drinks milk.

Snakes suck the milk out of cows.

This tale was often told in relation to the Red-bellied Black Snake. Whether it evolved as a simple justification for clobbering snakes or an excuse for a poor milker that the farmer was reluctant to send to the knackery is anyone's guess, but snakes can't suck cows. I've watched a thirsty snake drinking water and they are very much uphill doing that. Their sucking mechanism is very low powered and it takes them ages to draw up a very small quantity of water.

A saucer of milk will attract a snake.

I first heard this one at primary school. I think it's mentioned in Henry Lawson's *The Drover's Wife*. Snakes are not attracted to milk. Thirsty snakes drink water.

Snakes can jump.

No, they can't. Jumping is about having legs. The most extreme form of this myth was told to me by an elderly lady who claimed that her uncle had died after being bitten on the neck by a Death Adder. "And," she told me, waggling her finger in my disbelieving face, "he was sitting on a horse!" I was on the verge of hysterics but I could tell she was serious.

Birds commit murder.

If we remember that there are no police services or criminal justice system in a wildlife community, but that "law and order" must somehow prevail, we quickly realise that a "police service" of sorts is built into each individual. In some invertebrates such as ants and termites there is an army (of sorts) comprising soldiers to protect the nests, but there are no organised groupings that deal with aberrant behaviour within the nest. There are no jails in which to confine "wrong-doers" and there is most definitely no concept of revenge, or "an eye for an eye". There is no revenge in Nature. What people have observed and mistaken for "murder", "theft", "adultery", and so on, is, almost without exception, perfectly natural behaviour that is misinterpreted because it is judged according to human behaviour and values. This is known as anthropomorphism and needs to be watched very carefully.

If one member of a family group of birds suddenly begins to behave abnormally because of illness or injury, there is a danger to the rest of the group that this behaviour may attract the unwelcome attention of predators. Many predators, the shark being a good example, are attracted to just such behaviour. Then everyone is at risk. The aberrant behaviour has to be either avoided or stopped for the good of the community. Non-territorial birds, such as lorikeets, can simply flee, leaving the dangerous behaviour behind them, and this they often do. But sometimes the flight option is not available and so the healthy birds attack the sick or injured until the abnormal behaviour is eliminated. This often means death.

Baby birds will be rejected by their mothers if they have been handled.

This is probably a very useful story to keep small boys away from birds' nests, but as the vast majority of birds have no sense of smell, there is no way for the mother or father bird to know it has been handled. The very real danger – and I have witnessed it more than once – is that handling by humans can betray the nest to predators that do have a sense of smell, especially the family cat.

The scent left on the bird and the nest when the baby is returned can cause the nest to be investigated by pet cat or dog and the nest destroyed. Interestingly, the scent of humans on possums, flying-foxes and Koalas appears not to affect maternal behaviour at all. Many times we have successfully reunited mum and babe after separation has been caused by sexually aggressive males or territorial disputes. So even wild animals with very good senses of smell don't seem to be fazed by the dreaded human scent.

Frogs' skins are "burnt" if handled with dry hands.

No, they aren't, but I suspect that the origin of this one is well founded. Native frogs' skin is covered with a protective mucus which will be removed if the frog is handled roughly. Wetting your hands probably reduces the amount of mucus rubbed off, but if the frog is well hydrated, the mucus is replaced very quickly. There is certainly no "burning" effect.

Bats are blind.

The term "blind as a bat" is a most unfortunate one. Although insectivorous bats augment their vision with echolocation, as do many cetaceans (whales and dolphins), all bats see quite well. Echolocation is the mechanism by which animals vocally transmit high pitched sounds and then interpret the echoes which bounce back. We often hear it referred to as radar and, although the principles are similar, it is not radio waves being transmitted but sound waves.

Bats fly into your hair.

This appears to be a very old story. It is said to have originated in the UK. One version is that a bat was chasing a moth which flew into a lady's pompadour hair-do and the bat got tangled up in the hair along with the moth. Who knows?

Thrill killing.

Many times I have heard people describe the wrecked chook run as having been the work of a fox and foxes kill for the fun of it. It is easy to come to this conclusion but it is wrong. I'm not defending foxes as serious feral pests in Australia, but I am still offended by the "thrill kill" notion. The same is said of Dingos and their attacks on sheep. In both cases the reason that all the chooks or sheep are killed rather than just the one the fox or dingo wants for the meal is that they are programmed to do one thing at a time.

When the fox climbs (yes, they climb very well) into the chook house at night (which is why I recommend a wire roof on the chicken run) it grabs the nearest chook. Now if the birds were wild ones rather than penned up ones, the rest would flee. The caught one would stop moving when dead and the fox would then proceed to the next step which is to eat it. The problem in the chook house is that the chicken isn't dead. It doesn't stop moving. There it is again. But it's not the original chook, it's another one, so the fox grabs that one and another till all the movement stops. Then it can move to the next step. "Thrill killing" is not known in Nature. The Dingo in the sheep pen is doing the same.

NB: Of course, if anyone is totally convinced that these things are true, they won't believe me if I tell them they aren't!

To feed or not to feed?

That is the question. The answer is not so simple and the whole issue has been muddied up by a combination of facts, furphies, half-truths and misunderstandings. Oddly enough, the anti-feeding stance is peculiar to Australian culture. Overseas, the deliberate feeding of wildlife, particularly of birds, is mostly organised, advocated and promoted by respected universities and their professors of ornithology. In Australia, it is generally frowned upon but recent research at Griffith University is showing that there is much to gain and little, if anything to lose.

The desire to be kind to the wildlife is very strong in some people and this often extends to providing handouts. On rare occasions, the provision of gradually more and more food has resulted in very large aggregations of birds, particularly magpies, which have eventually led to neighbourly disputes over soiled washing as dozens of birds perch anywhere they can, including on the neighbour's *Eucalyptus hills-hoistii*.

Some of my friends enjoying Devonshire Tea.

But exceptions neither make the rule nor are the rule, and in general backyard wildlife feeding is a simple pleasure that can be carried out in comfort and safety. Moderation, as in most things, is probably the best guide.

Debates on this issue often end up being about nutrition, which is largely irrelevant. There is a world of difference between the bird in the cage and the bird that flits in for a handout. Some of the claims made against feeding are wilder than the birds themselves. The food is blamed for a whole range of well-known diseases that are caused by viruses and bacteria, not nutritional deficiency, and some cases, while having a sound theoretical base, are constructed to fit the dogma.

Occasional or regular offerings to the wildlife in your back yard will do no harm and is your affair. Outside your back yard, it is a different matter.

Q. *We feed the birds in the back yard. Someone told us this was a bad idea. Am I doing the wrong thing?*

A. No. Research by Dr Darryl Jones, his colleagues and students at Griffith University has shown that no matter how much we try to feed the backyard wildlife, we will never be providing more than a small proportion of their daily food requirements.

In these studies, magpies were provided with enormous amounts of food, but only 16% of the total of what they ate daily was the mince and bread that was put out by humans. The rest was what they foraged for themselves.

Q. *We were told they would become completely dependent on us. Is this true?*

A. No. Even if the birds were taking a lot more than the 16% that the studies show, birds are incredibly resourceful animals. It is the essence of their success.

If you go away and the food you were providing is no longer available, they will simply increase their foraging. This is logical. I have seen a huge Moreton Bay Fig Tree absolutely dripping with fruit. In this tree was a spectacular show of fruit-eating birds. Two Channel-billed Cuckoos, a Common Koel and about a dozen Figbirds could be made out bobbing about in the branches above. If the dependency theory were true, what were they to do when the figs ran out? If food can't be found in one place, birds move on. They don't hang around the depleted tree and starve.

Q. ***Our magpies, butcherbirds and peewees just love cheese. We were told cheese was bad for them. Is this true?***

A. No. Probably not. At 16% of the total diet, it probably doesn't matter much what you put out for the birds.

We should remember that we do this to please us rather than as a necessity for them. The claim that cheese is not a natural food for birds is certainly true, but then meat pies aren't a natural food for crows but they eat them and look all right to me. I have been contacted by people who have fed cheese to their local birds for years, know them all by name, and it seems they are living a healthy, normal life span.

This Rainbow Lorikeet is totally ignoring the feeder tray (not visible in photo) just below this flowering Red Ash.

Q. ***We found a website which warns of the dangers of feeding lorikeets. We've been feeding the lorikeets in our back yard. Should we stop?***

A. No.

I have seen this website and wrote off for further information. The site's owner was unable to substantiate the claims made and I concluded that the "fact sheet" was a well meant fabrication. The only properly published results on lorikeet deaths that I can find, relate to three lorikeets that died in a pet shop in Victoria. Again, in the captive situation the feeding rules are radically different.

***Q.** We'd like to feed the birds in our back yard. If we do, what should we feed them?*

A. You can offer different foods to different groups, but don't be surprised if you find the magpies getting stuck into the parrot seed. I have made suggestions here for the different food groups.

For the meat-eaters – magpies, butcherbirds, drongos, kookaburras, peewees – will invariably go for strips of lean meat, especially heart. Because what you are feeding them is such a small proportion of their daily diet, "anything they will eat, including cheese" is probably the correct answer, but, as the feeding of wildlife has become such a controversial issue, I think it's best to avoid criticism by putting out food that even the vet would recommend. So, if in doubt, give your vet a call.

For the big seed-eaters – including cockatoos, galahs, corellas, large and small parrots, rosellas and lorikeets – I don't think you can go past grey-striped sunflower seeds. I was told by a vet that the completely black oil-seed form is less suitable due to its very high oil content, so I've always stuck to the grey-striped form.

For the small seed-eaters – including pigeons, doves and finches – choose a good quality budgie mix. It's probably only going to provide 16% of their daily needs, but why not go for a good quality mix?

For the nectar feeders – including lorikeets and honeyeaters – mix one part honey with nine parts water. The lorikeets will love it. If you want to include some wholegrain bread, fine. There have been all sorts of warnings of the dire consequences of putting out honey and water so I recommend that you thoroughly wash the feeding bowls daily and only put nectar mix out for short periods, say an hour or so.

The birds will get used to your timings and turn up on cue. When the mix is gone or the birds have got their 16%, they will depart. Clean up. There's no point in feeding the wildlife if you're not going to watch. That's what it's all about. They don't need your help. You need theirs.

Alan and Stacey Franks

Seed-eaters with bird-feeders.

Q. *We were told that feeding the cockatoos would cause beak and feather disease. Is this true?*

A. No. Beak and feather disease is caused by a virus.

Psittacine (derived from "psittakos", the Greek word for "parrot") beak and feather disease, more properly known as Psittacine Circovirus Disease, is caused by a virus shed in the droppings of infected birds rather than by eating sunflower seeds. In the wild, it is a serious and tragic problem for the individual bird, but not for the species as only small numbers of infected birds succumb and there are no records of whole flocks being wiped out. It is most noticeable in Sulphur-crested Cockatoos; affected birds fail to replace lost feathers and there is often abnormal growth of the beak with the tips elongating and curling as our fingernails do when they are left uncut for a couple of years. As the cause is viral, there is no cure.

One argument asserts that deliberate feeding will tend to concentrate birds in a small area and increase the risk of infection. The trouble with this view is that cockatoos always feed in flocks. It is very much a part of their safe-feeding strategy.

There is always one bird with an eye out for predators. This is why the lookout at an illegal two-up game or SP bookie's operation is called the cockatoo.

The real issue is that infection can be caused anywhere fresh droppings can be ingested, so strict hygiene at your feeding stations should prevent criticism.

Sulphur-crested Cockatoo: a Psittacine Circovirus victim.

Q. We have possums in our yard. What do we feed them?

A. Fruit, vegetables and cereals are best.

Brushtail and ringtail possums will eat a whole range of fruits and vegetables. The best approach is to start with the peelings of whatever you had that day, such as apple, potato and carrot peel with a bit of sliced banana. Then just let them teach you. With some notable differences, brushtails and ringtails are predominantly leaf-eaters so you can expect that, like the birds, they will probably be only taking a small proportion of food put out by you. I've known healthy brushtail possums that for years have been given a small piece of fruit cake every night as they come out to forage. They will eat a little to please you but if you put out a large amount it will just be left.

They're smorgasborders. It's probably a great strategy to avoid intoxication as most of the native plants they eat have some nasty chemical designed to protect the plant. In small doses these toxins have little effect, but the smart possum (and they all are) nibbles a bit and moves on.

Q. I saw a program on television where a bloke was feeding dozens of Squirrel Gliders with just white bread and honey. Isn't that really bad?

A. No, probably not.

I saw the same program and I know one case where this was happening for years with no apparent ill-effects. I think it's probably the 16% factor again. While it looked as though the gliders were eating enormous amounts of honey and bread, in fact the total amount eaten by each glider was probably only a small part of what was eaten that night. Either that or the bread and honey was good tucker so far as the gliders were concerned. After all, isn't the proof of the pudding in the eating?

Steve Parish

Squirrel Glider.

Q. ***We've had heat wave conditions here and last night we found a Koala drinking from the dog's water bowl. Should we put out some water?***

A. Yes. Definitely yes!

During prolonged periods of above average temperatures, heat stress can cause the deaths of a large number of native animals.

There's not a lot we can do about that in the bush and it is most likely that population numbers will be sufficient to sustain a reasonable level of mortality. But, in urban environments and the surrounding areas, there are already serious impacts on wildlife numbers through habitat loss and fragmentation, roads and domestic pets, so a bit of extra help during a heatwave is a sensible strategy to reduce further mortality.

Put the water in a heavy bowl to provide stability and make sure it is thoroughly cleaned each day. Oh, and please make sure the dog is not a threat to Koalas (or any other wildlife) at night. If your dog weighs more than 7 or 8 kilograms, it is physically capable of mauling and killing a Koala regardless of its breed and temperament. Our dog stays in the house at night to keep the Koalas safe, and, if you are concerned about unwelcome human intrusion, with all of you locked safely inside where the phone is, who can do you harm?

Steve Parish

A Koala on the ground is vulnerable to attack by pet dogs.

Domestic poultry deliberately installed in a council park. A bad idea.

Q. *At the local lagoon, people feed the ducks. Should we do that?*

A. **No, definitely not!**

Firstly, it's not your back yard and that's the cardinal rule of wildlife feeding. Secondly, the "ducks" are probably a combination of wild native ducks and a collection of barnyard birds, such as Muscovy and various Mallard breeds. I've even seen Greylag Geese (those huge grey and white birds bred for eggs and the table) repeatedly installed in a large ornamental lagoon in south-east Queensland. Each time the local authority, with the help of the RSPCA, rounded them up and found them new homes, within a few months more birds had been dumped there.

Tossing bread into water in someone else's back yard is not something I would ever do despite my otherwise very liberal attitudes. It all gets unbelievably messy and my wildlife feeding program definitely includes good hygiene. Inevitably, everyone expects someone else to do the cleaning of the local lagoon. If the native ducks come to your back yard – that's different.

Q. *We're on acreage and we see wallabies and kangaroos. Can we feed them?*

A. **Certainly, and the best way may well be to leave the mowing to them.**

Use some of the surplus dam water to sprinkle the grass in dry times to keep up the food supply. If the supply of water is tight, get some good lucerne hay and put a bit out in the late afternoon and see if you can get some "roo pellets". There is a specially made macropod (kangaroos and wallabies) food that is made for zoos and research facilities specifically to feed these animals.

You can also get a medicated version designed to prevent a potentially fatal disease known as coccidiosis. If you are putting food out, put it in a different place every few days to prevent a build-up of the microbe which causes coccidiosis. Ask your local produce store to get in some roo pellets, but consider the grass option first, especially locally indigenous grasses. These are becoming much easier to obtain from native plant nurseries.

Q. *Our wallabies seem to like a slice of bread or two. We were told this would give them something called "Lumpy Jaw". Is this true?*

A. Essentially, no, because Lumpy Jaw is caused by bacterial infection.

A range of bacteria have been identified in cases of Lumpy Jaw, but the primary culprit appears to be the bacterium *Fusobacterium necrophorum*. This organism lives in the soil and is very common. It can invade the jawbone via the tooth root as a result of injured gums or gums badly affected by gingivitis.

In captive macropods, prolonged feeding of very soft foods, such as bread, can cause gingivitis and allow the invasion of the bacterium. Equally, however, a diet which includes hard, sharp pieces can also injure gums. Where the bacterium is present in high concentrations, infection is more likely. Put out a slice of bread, plus some roo pellets, plus a sheaf of lucerne hay and the recommended diet is on track. Although the chance of free-ranging wild macropods developing Lumpy Jaw is fairly remote, avoid a bacterial build-up by regularly moving the feeding site.

Q. *We left some fruit out on the veranda last night and a flying-fox came in for it. Can we feed them?*

A. Yes, why not?

Tie a bowl in a tree and fill it with a combination of chopped pawpaw, apple, banana and, if there's a mango glut, some mango too. There's nothing wrong with putting out a bit of fruit to get a closer look at these native animals. They'll keep you entertained for hours if you let them. There aren't too many programs on television that can compare with the antics of the Big Bats.

Lyssavirus – a relative of rabies – is a reality now in Australia, but, one has to be bitten or scratched by an **infected** animal to risk developing the disease. Keep your distance. There is little risk from cleaning the fruit bowl next morning, but the wise don't handle the uneaten fruit with bare hands. To be absolutely safe – wear rubber gloves. (Lyssavirus needs to be taken very seriously but without panic. If the disease is contracted and does develop, as with rabies it is always fatal.)

Q. *By feeding the flying-foxes commercially produced fruit, won't I be teaching them to raid orchards and cause trouble for them and the growers?*

A. No. Flying-foxes find food by smell and probably by cues given off by their camp cohorts.

Unfortunately they do not need to be taught to find orchards. Unprotected crops will always be raided by some flying-foxes and all we can hope is that one day very soon there will be no unprotected orchards. Netting works wonders.

Babes from the woods

As I am writing this, I can hear the incessant "chip... chip... chip... chip" of EPIRB, the baby Noisy Miner, through the open window. EPIRB came to us as an orphan. We can't justify euthanasing healthy native animals just because of some misfortune, so we began to feed it in a makeshift nest in a cage on the back veranda. Within a very short time, our local miner group took an intense interest in EPIRB and tried to stuff food into him through the wire. They had a youngster a month or so back but we found it dead in the yard one morning. Who knows why? Their interest in EPIRB was so intense we decided to let them look after him and that is what they have done now for a week. It looks as though EPIRB is going to be a success where their own was not.

Are we doing the right thing? I think so. Firstly we're not speciesists, so we are unconcerned about the fact that it is a member of an abundant native animal species. Should EPIRB have been left to die an orphan? What if no one had found him, or bothered to rescue him?

These questions have rattled around for a long time and I'm not certain there are any definitive answers. I have heard people express views that they know what's best and that all tragedies in the wild should be left for Nature to take its course.

After some years of pondering, I came to the conclusion that what you can't know can't concern you, but what you do know does. If a tree falls in the forest and I'm there, I will hear it.

Orphaned baby Wood Ducks – all thrived and survived.

Over the past 25 years I have shared my home and family with an enormous array of wild animals, including nearly thirty orphaned Koalas, at least thirty species of birds, and a few very intriguing characters including a baby Water-rat that had a growl and a bite that could put the wind up even the most fearless Wildlife Ranger. Most recently we tried valiantly to raise a baby Platypus. We failed, but the post-mortem suggested what we might do to give the next one (should that ever happen) a better chance.

Baby platypus – we did everything we could.

I started in the early 1980s with kingfishers. I remember the first one and I remember the day it happened very clearly. I was driving along a busy road not far from the Central Business District of Brisbane when a bright aquamarine flash did a wheels-up crash-landing on the road and slid into the gutter. I had to stop. It was a Sacred Kingfisher.

It was exhausted and easily captured. There is just something about a real wild bird in the hand.

He (or she) got the name Eddie (Eddie Fisher) and I got in touch with the wildlife carer's group via the Queensland Museum to find out a bit about caring for Eddie. Sacred Kingfishers can lay quite large egg clutches, often five or six, and the parents become a bit less concerned about the "Tail-end Charlies". If you've raised the first three or four of the offspring you've done a sterling job, considering you've flown all the way from Papua New Guinea to do it. Eddie proved to be a totally uncomplicated, straightforward case of a "Tail-end Charlie".

I simply fed Eddie five or six big feeds a day. It was case of open beak forcibly, pop diced beef heart coated in DCP powder (there are much better diets now) over the tongue and not down the windpipe, followed by a few drops of water, and back on the perch.

Eddie flew all over the house, woopsied on everything, but that could be cleaned up in a flash, and flew off with not a word of thanks a couple of weeks later. I treasure the pictures I still have of him. I was a furniture designer/maker at the time, but life was soon to take a very different turn.

I was very privileged to assist with the hand-raising of a Yellow-footed Rock-wallaby that had accidentally been separated from its mother during a research project. There is nothing more startling than to have a couple of kilograms of tiny wallaby come bounding across the lounge room floor and, in a single leap, land on your shoulder and stick its nose in your ear. Rock-wallabies have broad, crepe-soled feet and no fear of heights.

It terrified me one morning when I saw it had got out of the room it had been allocated and was perched unconcerned on the edge of the bathroom window looking down some eight metres to the ground below. There was no event. It simply spun around and dropped to the floor inside the house and headed back to the "pouch".

Wildlife welfare is not the sole province of silly, soppy animal lovers as some might have you think either.

Steve Parish

Yellow-footed Rock-wallaby.

Orphaned by a motor car.

A friend of mine, Dr Frank Carrick, is a senior lecturer in biology at the University of Queensland (UQ). He is also founder and chief investigator at the UQ Koala Study Program. This program owns and operates a small but very significant Koala Hospital which rehabilitates orphaned, sick and injured Koalas for release to the wild.

Frank doesn't seem to have any difficulties in justifying his hospital's existence. True, the hospital has always doubled as an important research facility as well, but there have been an awful lot of patients through it over the years that were never subjected to any research and simply got high quality treatment for their illness or injury and, so far as the Koalas know, Frank must bulk bill all his patients.

During the 1990s, the UQ Koala Hospital, in conjunction with the nearby Queensland Parks and Wildlife Service's Moggill Koala Hospital, returned an average of about 200 Koalas to the wild that, in former times, would simply have died. Two hundred per year adds up to 2000 in ten years and that's a viable population.

The Koalas treated at these hospitals come largely from the last remaining urban and peri-urban population from the very last Australian capital city to still have wild Koalas. Sydney, Melbourne and Adelaide all lost theirs some time back. Until very recently the Koala was not considered to be a species at risk in Queensland but, despite this, a lot of effort was put into caring for the orphaned, sick and injured individuals.

The recent announcement that the Koala will be officially listed as a vulnerable species in Queensland suggests that the research and welfare program begun in the early 1980s by Frank Carrick at the University of Queensland was far-sighted and well ahead of its time.

Koala population surveys that were being conducted in south-east Queensland in the mid 1990s were drastically underestimating the number that was out there. We only knew this because we had an extensive capture, mark, release and recapture program in operation, except it wasn't called that, it was called the Koala Hospital. All Koalas that came through the hospital were ear-tagged and micro-chipped before being released.

Computerised records of these animals included size, weight, sex, head length, body length and tooth wear (or age class) which told us a lot when, inevitably, some got into trouble and were returned. Illness, injury and "stranding" in back yards all worked in our favour. Independently, scientists were combing the bushland counting Koalas in trees and extrapolating the results to the total bushland.

Preparing for a life in the trees.

Their estimates just did not add up to the numbers that had to be out there. A correction in their technique eventually brought the two estimates within a hundred or so of each other.

Had the Koala Hospitals not existed, the field search estimates may have underestimated the population to such an extent that efforts to conserve them might have been abandoned as a hopeless case.

A detailed analysis of more than 200 orphans that eventually returned to the hospitals over more than a 15-year span showed that hand-raising was no barrier to a productive life in the wild, despite all sorts of fears expressed to the contrary.

One such former orphan that is presently thriving in the wild was only 56 grams when he lost his mother!

Much is often made about the necessity to provide natural foods and to mimic the mother's training techniques for the orphan to survive. There are a few species to which this applies, and, in those special cases, it is extremely important to do just that, but my experience is that most species are fairly flexible. Birds and bats are not taught to fly. They just do it. Try to stop them! Flying-foxes will go from commercially produced human milk formula and orchard fruits to their wild foods without any hesitation. Possums, gliders, kangaroos and wallabies all do the same. The special cases include Koalas and Greater Gliders which must be weaned to a diet of eucalypt leaves; birds of prey which must be specially prepared and trained to hunt; and birds, such as ducks, which are confined to the nest for a time after hatching and which can be mal-imprinted. They do require specific expertise. But most wild animals will do well, provided they are healthy on release and provided the release technique is not just a case of dumping them out there.

If the endangered Bilby is ever to come off that unhappy list it will probably only be by re-establishing viable and expanding populations through intensive breeding and releases in suitable areas after appropriate research has established the cause(s) of the original extinction. The expertise to accomplish this is derived directly from captive management techniques originally pioneered by welfare programs.

There is another side to all this as well. Some people don't want traditional pets in their lives, but do want some sort of contact and responsibility for an animal or animals. They don't wish to try to make native animals into pets (a bad idea – see Pet Theory of Conservation on page 115) but are keen to have some sort of close contact. A wildlife welfare program provides this, at the same time producing more allies to the conservation cause.

I have many times seen blind prejudice reversed by a close encounter. Baby flying-foxes are superb at it. One look from their huge intelligent brown eyes and you'd have to have a heart of granite not to be moved.

Those who view wildlife welfare programs as little more than "nurturing neuroses" which contribute nothing to conservation are often heard to say that they are more concerned with The Big Picture. I always like to remind them that a big picture is made up of a lot of little dots. Ignore all the little dots and the big picture will disappear.

Then there are the murmurs that a lot of the animals that are rescued and cared for were not meant to survive and doing so goes against the natural order of things, ensuring the survival of individuals that could go on to muck up the genetics of their species. Does anyone have any evidence to support this? Not that I know of.

Land use practices in Australia over the past 200 years or so have so badly fragmented the natural world that roads, suburbs and farmland will cause more genetic decay in less mobile species than the welfare program will ever do.

Equally, people are concerned that hand-raised wild animals often fail to survive when released. There are a few observations and studies that actually demonstrate that this is indeed the case. The only problem with these studies is that they tend to be looking at the world through a drinking straw. If only one in twenty hand-raised baby magpies survives after release, this is an excellent result because only one in twenty baby magpies raised by their parents survive anyway. We have to compare the results with Nature, not with our own expectations of life. My observations suggest that hand-raised magpies, for example, actually have slightly better survival rates than their wild-raised counterparts.

Welfare programs can also be the source of remarkably valuable data if the effort is made to record, store and retrieve the information. Tens of thousands of native animals come into care every year. An awful lot of carers keep basic records of these cases.

If even the date, species and location only were recorded and loaded into a central database, a few years' worth of data would be a veritable goldmine of information.

Not all of the animals that come into care are well known and abundant species either. A little black fluffy ball on legs handed to me grew into Lewin's Rail, a species I have still to see in the wild.

A Lewin's Rail – the only one I've ever seen.

Few people know that the rediscovery of the Northern Hairy-nosed Wombat, once thought extinct and still critically endangered, happened because one was successfully raised in captivity after it was orphaned. And it was still there in 1974 when the zoologist who had been told of its existence went to see for himself.

Conservation programs probably work best when they have a number of strings to their bow, and an active and successful welfare program is a great little public relations exercise when you are trying to convince communities, developers and governments to set aside quality habitat as nature reserves.

If you happen to be looking for a good photograph of a native animal and there aren't any in the archives, or if you are looking for a good story to do for television, call the carer network. There'll be something there, somewhere. The endearing cover photo on the best selling book, *Wildlife of Greater Brisbane,* is of two Tawny Frogmouths that were in care, being hand-raised at the time that the picture was taken.

People who find wildlife welfare work all rather tiresome should simply leave it to those who aren't the slightest bit tired.

Anyway, if you're ever unlucky enough to be driving past a dead female Koala that has a little furry joey clinging forlornly to its deceased mum, just look straight ahead and keep driving. Just try.

Orphaned by people "tidying up" a "dangerous" tree.

Birds in big buildings

The big birds

Kookaburras, magpies, butcherbirds, sparrowhawks, goshawks, frogmouths and owls have all been reported as having got into a very large building and found themselves seemingly unable to get out. The problem has several facets. Sometimes the building's owner is concerned about loss of productivity. It's a warehouse or factory and the workers are spending too much time being either concerned or amused by the bird's "flappulence". At other times the concern is the same as mine – that the bird will eventually starve. We know from an unintentional experiment on a Laughing Kookaburra in a television studio in Brisbane many years ago, that it takes about one week for a largish bird to starve to the point that it can no longer perch high up and falls to the floor.

Emma Ralston

Wild birds often "stumble" into buildings like this.

ew people would want to run the risks associated with this method of dealing with the roblem and quite reasonably want a quicker solution.

remember my first attempt. A Laughing Kookaburra had flown into one of the biggest ndustrial buildings I'd ever stepped into. The roof was about three storeys high. Steel eams at regular intervals across the vast expanse formed the bottom of the triangular nonstrosities that held up the roof.

The kookaburra could fly the length of the building through the gaps in the structural steel, and at the other end and laugh. At me. I spent some time pondering the bird, the building nd chances of catching it. There were plenty of doorways for it to fly out but the doorways were only about one storey high and the bird didn't seem to "realise" that it needed to dip down to fly out. It is likely that the presence of humans working below was the factor that topped it from doing this. This bird had been in the building for a few days before being eported so its welfare was of some concern.

had a catching net strapped to the end of an eight-metre-long pole and with it I danced up nd down the warehouse, dodging the crossbeams and hoping that the kookie would make mistake and fly into my tiny net. My assistant was one of the workers who chased the bird rom one end of the building to the other using another pole with a flag on its end. It was ike upside down reverse badminton with oversized rackets and a shuttlecock that defied all he normal laws of physics.

Miraculously, about an hour and half into the game and after a lot of rest breaks, the bird did make the desired mistake and flew straight into the net. A cheer went up from the workers below and there was much back-slapping and congratulations all round but, for me, it was a complete shambles. There had to be a better way.

There is and there are. Many big birds in big buildings later, I found that three methods can be used depending on the particular circumstances prevailing.

The skylight method

Some large buildings have skylight panels in the roof. Someone who is not afraid of heights (I am – very) can get up on the roof and open the skylight or, if there is no skylight, one can be installed temporarily by removing one panel of the roof. This might sound like a lot of trouble but in some circumstances it is easiest and cheapest. The bird will soon discover the opening and whiz out. It can be assisted in this by being chased around a bit.

The best "chasers" are those huge surf rods with a hankie tied on the end. This is one of those occasions when the rod can be used for good rather than evil (so far as the fish are concerned).

The dinner-by-the-door method

Most big birds I've seen in big buildings are of the animalivorous (live food) type. I've used a mouse in a cage on the floor by the door to attract a kestrel and butcherbirds down to ground height. When the birds realise they can't get the mouse they go for option two and shoot straight out the door. Pieces of cheese or meat will work on magpies and anything that will move in a big wide plastic dish (cockroaches are good) will work provided the bird is more hungry than wary.

The Chase-the-Cheetah method

This is by far the best method if circumstances permit. It involves chasing the bird to exhaustion in one single go. It is important that the bird is not permitted to rest even for a second. Several chasers with surf rods wave these flags at the bird and keep it on the move continuously. Have a big towel or sheet ready to throw over the exhausted bird. In a very short time the bird's wind will be blown and it will flutter down the wall to safe capture. Put the bird immediately into a completely dark cardboard box and allow it to recover.

This sounds terrible, but it isn't if it's done in a very short period. It needs to be stressed that if the bird is allowed to rest and partly recover to be chased again and again, great damage will be done. Chase the Cheetah is about rapid "blowing", just as the Cheetah does when it chases the gazelle. If the Cheetah's chase fails (as so many of us who've watched those Big Cat shows on television always hope it will), it pulls up totally winded and needs to rest for a time before it can do it all again. This is the principal that is being used here. I've done it dozens of times and when the recovered bird (after an hour or so in the box) is released it always tears off into the sky looking in great shape. Often it's either that or starve to death in the building.

Little birds in big buildings

Swallows and martins like big buildings. They are related to birds that like cave entrances and big amphitheatre-like cliffs. They roost on them and build their nests on them. In the absence of real caves, large warehouses and factories with huge doorways will do, and the owners of these building can find themselves host to nesting and roosting swallows and martins. If the inside of the building is a mass of steel beams, ledges and braces, the chances (not to mention the expense) of altering the structures to make perching and nesting impossible are usually slim.

The objection to the birds is fairly straightforward. The birds drop things and this offends hygiene and workplace health and safety standards. These buildings usually have great big doorways through which go forklifts and other vehicles as well as pedestrians and, of course, the little birds. There is a great benefit for the birds. Inside the "cave" they are safe from predators such as the Little Falcon (Australian Hobby) which is very partial to a bit of a swallow of swallow. Chasing the birds around the building is a waste of time. They instantly know you can't get them and although I've been told of a whole host of scary devices that will shoo swallows away, none of them has worked for me beyond the first few days.

What does work is the proverbial barrier. The best barrier is clear plastic sheeting, cut in broad overlapping strips the way it is used attached to the doors of cold rooms. People can see through it, so that drivers of vehicles don't have unnecessary contact with each other and the birds can't fly though it. It may sound like a lot of trouble to go to, but if keeping the birds out is an imperative, it's the way to go.

Pet theory of conservation

There's a theory that says: if we all kept a native animal as a pet, rather than a dog or a cat, Australia's conservation problems would be significantly reduced. The arguments contained within the proposal are initially extremely attractive. Should they stray or wander off, not to worry – they're native to Australia, and they won't cause the sorts of problems the feral cats and wild dogs have. Even better, if we kept rare and threatened species as pets it would be a huge bonus, because we would have insurance against their extinction. If we made pets of, for example, bilbies, then we'd always have bilbies and if they went missing in the wild, we would always be able to put them back using some of the pets to breed up stock for release.

Some of the most vocal proponents of this theory were, until fairly recently, some of my greatest conservation heroes. Then they decided to champion the "pet theory" and I had to sack them. But we need to clear up one huge misconception.

The "pet theory" is not really advocating native animals as pets. There are hundreds of species of native animals already in captivity. You can buy and keep a taipan tomorrow with the proper permits. They are talking about native mammals which are not allowed to be kept as pets in most States of Australia.

Pet Koalas, pet rock-wallabies, pet quolls are all candidates, and nothing sounds better than the contact part of the theory which says that Australians would have a much better understanding of and a greater concern for the conservation of their wild species if only they were able to relate to them. How can people be concerned about the long term survival of Mulgaras if they have no idea what a Mulgara is?

It all sounds very convincing, until you scratch beneath the surface.

For the "pet theory" to work, you have to make the keeping of a native animal as a pet as attractive as its argument. It has to be widespread. To make it widespread it has to be simple. The last thing you'd need would be a massively complex set of rules and regulations. You also have to be certain that the species to be conserved by this method actually make attractive pets. I've heard advocates talk about how incredibly popular Australian Sugar Gliders are as pets in the United States. Are they? Of all pet owners in the US, how many have Sugar Gliders?

In South Australia, where the restrictions on keeping native mammals as pets were lifted some years ago, what proportion of all pets are native mammals now? Dogs and cats are still the top two.I have also heard proponents try to negate arguments by theorising on the regulations and restriction to protect the wild populations and have been flabbergasted by their seeming lack of understanding of what a pet actually is. A pet relates to you. It comes when it's called. It isn't an unresponsive beast locked in a cage to be gawked at as a curiosity (and if everyone's got them they are no longer curiosities).

For pets to be responsive and safe you need firstly to strip out the genes that make them essentially wild. This is what we've done with the wolves with varying degrees of success. The failures are the ones that live in the yard with the Declared Dangerous Dog sign on the front fence.

So you manipulate the genetics to make them tractable. You mate sister with brother, carefully choosing the ones that don't bite, or at least you start with the ones that don't bite a lot. Then by selecting the wimpiest offspring you eventually end up with fairly trustworthy pets. We live with three dogs. All are genetically modified forms of the Grey Wolf with bits and pieces of additions. From a Grey Wolf conservation program perspective they are all idiots. We love them but they wouldn't be much use in trying to restore the world's Grey Wolf populations, which is exactly what some biologists are trying to do.

Toby Nattrass – a genetic idiot but a great pet.

Yes, all the world's wolf species and subspecies are in trouble! That's strange. The theory is that if people are able to relate to wild animals through their pets, the conservation of the wild species will be much easier. So what went wrong? To begin with, just about every mammal species we have domesticated has from then onwards fared rather worse than it did before domestication.

It appears that the wild ancestor of the goat disappeared so long ago that no one is really certain what it was. The Red Sheep, wild ancestor of all domestic breeds, is endangered. All wild cats are threatened, despite the millions of moggies in households around the world. The original horse is gone. Przwalski's Horse isn't it – there was an earlier version. The last of the European Cattle, *Bos primigenius,* apparently died in a Czech Nature Reserve in the 1670s.

Domestication actually threatens wild populations. The domestic form, more attractive than the wild one both in looks and temperament, becomes the prized version. Wild relatives have been deliberately exterminated, either to prevent their contaminating the improved version, or because the habits of the wild ones conflict with our requirements.

The fact that various domestic forms of the Grey Wolf were bred and trained as Wolf Hounds to hunt to extinction their wild ancestors should be enough of that part of the argument. With the conflict we see between agribusiness and the big kangaroos, it wouldn't take long to manufacture some pink-eyed, angora fleeced kangaroos and get rid of the pest ones!

I have been assured by one of Australia's leading ornithologists and bird artists that not one budgie cage he has looked into contained a single specimen of the real wild budgie. Even those with very similar colours were not the true wild form. So it seems that even if we don't deliberately wipe them out in the wild, we wipe them out in captivity.

I once saw highly desirable and therefore extremely expensive Gouldian Finches that had had all their irritating wildness expunged. Gone was the brilliant lilac breast, black, red and gold faces, canary yellow tummies and emerald green wings to be replaced all over by a rather grubby off-white. Thousands of dollars a pair! We just can't help ourselves.

Ian Morris

Gouldian Finches: the Wild Look...

...and the "improved" version.

And finally, if all the backyard cats were replaced tomorrow with backyard quolls, how would that help to reduce the domestic cat problem? Quolls weren't once known as Native Cats for nothing, and with male Spotted-tailed Quolls reaching six or seven kilograms I'd hate to be a bird or a lizard in their back yard.

We've domesticated enough of our companions and we do them no favours in the process. Rag doll cats too floppy to chase birds and tiny little pooches too small to even tackle a Blue-tongued Lizard are the way to go. We've already done it to them.

Cloud Cuckoo Land

Why we use the term "cuckoo" to describe something that's insane or whacky is quite beyond me. The brilliance of slipping your babies into someone else's nursery to have them raised at no cost to yourself is probably the sanest thing any parent could do. In the world of Nature the plot is to pass on your genes.

If you can get someone else to raise the kids, you've got time to pass on your genes, your T-shirt, your thongs and your sunglasses. It is a very smart strategy. Australia is home to about a dozen species of cuckoos.

The real odd-bod is the Pheasant Coucal or, less appropriately, the Swamp Pheasant. This "cuckoo" is a permanent resident that rears its own chicks, which the other cuckoos reckon is nutty. All the others are migratory and lay their eggs in other people's nests.

Steve Parish

The Pheasant Coucal, the only cuckoo to nest and tend its own chicks.

In eastern Australia there are also two big, loud, conspicuous cuckoos that arrive from the north in spring and make real nuisances of themselves. They are the Common Koel and the Channel-billed Cuckoo. I love them. Neither bird conjures up the typically cuckoo image and, since we are still largely suffering from undue European influence, we are always surprised when none of our Australian cuckoos go "cuck-oo".

The koel has a piercing, upwardly inflected "whoooeee" and a rising "weeoo weeoo weeoo" that sounds nothing like "cuckoo". It arrives earlier than the Channel-bill and calls all night long driving some urban humans to seek refuge in a Home for the Bewildered.

The Brisbane suburb of Toowong is named from one of the indigenous words for the koel and Too-wong can be sung very koel-like when you know how.

The male does the calling to let the females know he's there, and together they case the place for existing active nests. Their favourite aunties and uncles for their kids are peewees, orioles and the larger honeyeaters such as friarbirds and the Blue-faced Honeyeater.

The problem for humans is that the song has a rather mournful sound (for some) and then there's the fact that they barely pause for breath. They are also known as Storm Birds and Rain Birds but these names resulted from the coincidence that their spring arrival is also a time of afternoon thunderstorms in many parts of their range. During the later stages of El Niño, the birds will be out there Too-wonging away with not a cloud in sight!

It has nothing to do with weather, it's to do with sex. He is yelling, "I'm here sweetie! Do you come here often? What do you think of the band?" Not "We're all gunna drown!" There are people who simply do not sleep at all in spring when their big fig tree is full of romantic koels.

He is a longish slender all shiny black bird with big red eyes. She has grey bands and blotches but the same build and the same red eyes. They are not small birds – they are almost crow size – but due to their being *aves non gratia* with the locals, they are usually very difficult to see while being very easy to hear. They hide because they are "abused" by the incumbents who know their home-invasion habits.

The Channel-billed Cuckoo is even bigger and louder but tends to call less often, so you can snooze between singing sessions. His "song" is a grating screech that is difficult to liken to anything other than a throttled rooster, even though I am confident that none of us really knows the sound of a throttled rooster. The channel-bill favours the nests of crows, magpies and currawongs. It is a big bird. It is a big bird with a very big beak.

Many people unfamiliar with the channel-bill who see it for the first time often describe the beak as looking like a toucan and that's not too ridiculous a comparison. Both of these giant cuckoos are fruit-feeders as adults so it surprises everyone when they hear that their babies are raised on the sort of things that crows and peewees eat. These are miracles of Nature.

It has always intrigued me that while the parasitised birds recognise the adult koels and channel-bills as trouble, once the egg is in the nest they care for it as if it were their own, even when their "baby" has a great big orange gob the like of which they have never seen before.

The commonest call to Wildlife Talkback is for an explanation of what the big pale bird is doing harassing a group of crows. It's begging for food and it will pursue the crows because it has been raised by them, and that is perfectly natural to it.

But help is on the way.

A recent "discovery" (it won't be new to indigenous Australians) is that the adults seem to return to nest sites in which they've off-loaded eggs.

At the end of the breeding season they appear to come back and pick up their kids. We hear a return match of their calling in late January and throughout February after there has been a fairly long quiet spell. There is a theory that this behaviour is designed to introduce the youngsters to the adult calls of their own species. After all, the chicks have spent all their conscious life so far with what will turn out to be complete strangers.

It seems that during the course of a single breeding season, some koel and channel-bill pairs may be producing a dozen or so offspring, the rearing of which, if they had to care for them all themselves, would be an impossible task.

Cuckoo? I don't think so. And if you are tempted to feel sorry for the hosts, we have to remember that the big bad cuckoos are all fruit-eaters. The effects of their feeding is to spread the seeds of big trees, and there are a lot of magpies, crows and currawongs that are grateful for a nice big tree to bung a nest in and shelter from a summer deluge.

Conclusion

Some of the views I hold are not popular as they contradict accepted wisdom. The problem is that a lot of accepted wisdom can be dogmatic belief unsupported by science. A few years ago I played host to a young man who had been sent to Brisbane by his wildlife management authority to investigate the way the very tricky magpie nesting season was being managed in south-east Queensland. As I was driving him back to his motel on the final day, I was remarking on how far I thought we had come in some aspects of wildlife management during the 16 or 17 years I had been involved. As an illustration I told him that when I was in my first year as a Wildlife Ranger, we were even telling people to hurl handfuls of mothballs up into the roof to evict the unwanted resident possum and added some hearty laughter to lighten my embarrassment. There was silence. I looked over to him as he said in stunned, wide-eyed disbelief, "Doesn't that work?" Then I realised how far we still had to go.

Biologists constantly remind us of our country's dreadful record over the past 200 years in failing to conserve the uniquely Australian biota. If this is to be improved, it will only happen as the result of a significant shift in the way we all live and think. Whether we are grazing a vast area of the Queensland Channel Country or perched on a suburban block in one of the large capital cities, what we do will matter in the long run.

Biologists also often remind us that we are unlikely to improve our conservation record if we have no knowledge or appreciation of what is "uniquely" Australian. It all starts in the back yard regardless of its size and it must include tolerance as well as appreciation.

That, I hope, is what Talking Wildlife tries to achieve: a combination of better understanding, better management techniques where conflict occurs and a larger proportion of tolerance than we've displayed in the past. There is a lot to be done and a lot more good quality science is needed.

Trapping and shooting may not have been the only cause or even the principle cause of the extinction of Australia's largest and most spectacular carnivorous marsupial – the Tasmanian Tiger – but it wouldn't have helped and it seems extraordinary now that we did those things at all. Better stock management combined with tolerance may have saved it. And were the Thylacine still around in appreciable numbers, the "pest" shooting of Red-necked Wallabies in Tasmania may not have to happen at all.

During the months it has taken to plan and write this modest book and gather the photos, two male Brush Turkeys have disappeared from our local area and our back yard. Both have been removed by an intolerant gardener or gardeners obsessed with law and order. First to go was Giblets, as we ironically but affectionately called him. He turned up one morning with his entire tail missing.

As a former Wildlife Ranger of 20 years experience, I know this could only happen one way. He was trapped and mishandled by an inexperienced turkey hater, escaping with his life but not his tail.

It happened at the end of the mounding season so he was ignored for a few months and his tail regrew. But as the new breeding season began it wasn't long before a second attempt was made on Giblets' life – and he disappeared forever. Shortly afterwards, to our great delight we noticed a hatchling scratching around in the understorey of our garden and watched his rapid growth to full size. It happens quickly. He got the name Sog (Son Of Giblets) and later the diminutive – Soggie. Then he too disappeared at the height of this most recent mounding season. For some, Nature is something to be battled and beaten and, in Giblet and Sog's cases, for the sake of a sea of pansies.

But there is also a very positive and hopeful side as well. Each week, the board fills with callers to Wildlife Talkback wanting to find out what it is they have seen or heard or just to share an observation that intrigued or delighted. As the years have passed, there has also been a significant increase in the percentage of callers phoning to add their support for the underdogs – the Brush Turkeys, the possums, the snakes, the flying-foxes and, very satisfyingly for me, the most useful and intriguing of all the Australian birds – the crows!

Ric Nattrass

www.drivingyouwild.com.au

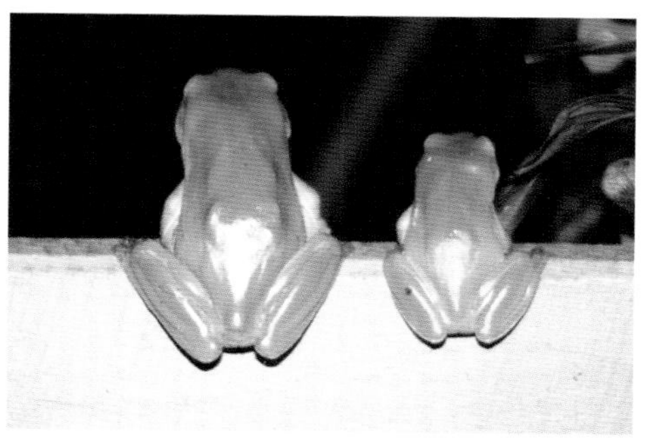

The End.

Great Books, Websites, CDs and Other References

Books

General

Encyclopedia of Australian Wildlife P. Slater, 2000, Steve Parish Publishing, Brisbane.

Handbook of Australian Wildlife C. Jones, (ed.), 2003, Steve Parish Publishing, Brisbane.

Wildlife of Greater Brisbane M. Ryan, (ed.), 1995, Queensland Museum, Brisbane.

Plants

Australian Native Plants J. Wrigley, 1996, 4th edn, Reed, Sydney.

Field Guide to Australian Wildflowers D. Greig, 1999, New Holland, Sydney.

Handbook of Australian Wildlife C. Jones, (ed.), 2003, Steve Parish Publishing, Brisbane.

Mangroves to Mountains Logan Branch SGAP 2002, CopyRight Publishing Co., Brisbane.

Wild Plants of Greater Brisbane M. Ryan, (ed.), 2003, Queensland Museum, Brisbane.

Insects and Spiders

Amazing Facts About Australian Insects and Spiders P. Slater, 1997, Steve Parish Publishing, Brisbane.

The Insects of Australia: a textbook for students and research workers (two volumes) I.D. Naumann, (ed.), 2000, 2nd edn, CSIRO Publishing, Melbourne.

Spiderwatch: A guide to Australian spiders B. Brunet, 2002, Reed New Holland, Sydney.

Reptiles and Amphibians

A Complete Guide of Reptiles of Australia S. Wilson & G.C. Grigg and M.J. Tyler, 1995, Surrey Beatty & Sons, Chipping Norton, New South Wales.

A Field Guide to Australian Frogs J. Barker, G.C. Grigg and M.J. Tyer, 1995, Surrey Beatty & Sons, Chipping Norton, New South Wales.

Amazing Facts About Australian Frogs and Reptiles P. Slater, 1997, Steve Parish Publishing, Brisbane.

Encyclopedia of Australian Animals: Frogs M.J. Tyler, 1992, Angus & Robertson, Sydney.

Encyclopedia of Australian Animals: Reptiles H. Ehmann, 1992, Angus & Robertson, Sydney.

Field Guide to the Frogs of Australia M. Robinson, 1993, Reed New Holland, Sydney.

Reptiles & Amphibians of Australia H.G. Cogger, 2000, 6th edn, Reed New Holland, Sydney

Birds

Amazing Facts About Australian Birds P. Slater, 1997, Steve Parish Publishing, Brisbane.

Encyclopedia of Australian Animals: Birds T.R. Lindsay, 1992, Angus & Robertson, Sydney.

Field Guide to Australian Birds M. Morcombe, 2003, Steve Parish Publishing, Brisbane.

Field Guide to the Birds of Australia K. Simpson (ed.), 1996, Viking, Melbourne.

Field Guide to the Birds of Australia G. Pizzey, 1997, Angus & Robertson, Sydney.

Magpie Alert – learning to live with a wild neighbour D. Jones, 2002, University of New South Wales Press, Sydney.

The Slater Field Guide to Australian Birds P. Slater, P. Slater & R. Slater, 2002, Reed New Holland, Sydney.

Mammals

Amazing Facts About Australian Mammals P. Slater, 1997, Steve Parish Publishing, Brisbane.

Encyclopedia of Australian Animals: Mammals R. Strahan, 1992, Angus & Robertson, Sydney.

Field Guide to Australian Mammals P. Menkhorst, 2001, Oxford University Press, Melbourne.

The Mammals of Australia R. Strahan (ed.), 2002, Reed New Holland, Sydney.

Scientific Papers

Rowley, I, 1969, 'An evaluation of predation by crows on young lambs', *CSIRO Wildlife Research*, vol. 14, pp. 153–179.

Rowley, I, 1970, 'Lamb predation in Australia: incidence, predisposing conditions, and the identification of wounds', *CSIRO Wildlife Research* 18, vol.15, pp. 79–123.

Websites

Australian Museum Online: **www.amonline.net.au**

Australian National Botanic Garden: **www.anbg.gov.au**

Birds Australia: **www.birdsaustralia.com.au**

Commonwealth Scientific and Industrial Research Organisation (CSIRO): **www.csiro.com.au**

Department of Conservation and Land Management (CALM), Western Australia: **www.calm.wa.gov.au**

Department for environment and heritage, South Australia: **www.environment.sa.gov.au**

Department of the Environment and Heritage: **www.ea.gov.au**

Driving you wild: **www.drivingyouwild.com.au**

Environment ACT, Australian Capital Territory: **www.environment.act.gov.au/bushparksandreserves**

Hollow Log Homes **www.hollowloghomes.com.au**

National Parks and Wildlife Service, New South Wales: **www.nationalparks.nsw.gov.au**

Parks Victoria: **www.parkweb.vic.gov.au**

Parks and Wildlife Commission, Northern Territory: **www.nt.gov.au/ipe/pwcnt**

Parks and Wildlife Service, Tasmania: **www.parks.tas.gov.au**

Queensland Parks and Wildlife Service, Queensland: **www.epa.qld.gov.au**

Western Wildlife (wildlife of Western Australia):**www.westernwildlife.com.au**

Index